Where Do You Go After You've Been To The Moon?

A Case Study of NASA's Pioneer Effort at Change

By Francis T. Hoban
with William M. Lawbaugh and Edward J. Hoffman

D0169171

AN ORBIT SERIES BOOK

KRIEGER PUBLISHING COMPANY
MALABAR, FLORIDA
1997

Excerpt from *Scuttle Your Ships Before Advancing: And Other Lessons from History on Leadership and Change for Today's Managers* by Richard A. Luecke. Copyright © 1993 by Richard A. Luecke. Reprinted by permission of Oxford University Press, Inc.

Original Edition 1997

Printed and Published by
KRIEGER PUBLISHING COMPANY
KRIEGER DRIVE
MALABAR, FLORIDA 32950

Library of Congress Cataloging-In-Publication Data

Hoban, Francis T.
      Where do you go after you've been to the moon? : a case study of NASA's pioneer effort at change / by Francis T. Hoban with William M. Lawbaugh and Edward J. Hoffman.
          p.   cm. — (An Orbit series book)
      Includes bibliographical references and index.
      ISBN 0-89464-060-7 (hardback : alk. paper).
      1. United States. National Aeronautics and Space Administration—Reorganization.   2. Astronautics—Government policy—United States. I. Lawbaugh, William M.   II. Hoffman, Edward J.
III. Title.   IV. Series.
TL521.312.H587   1997
387.8'0973—dc20   96-46071
                                                                              CIP

10   9   8   7   6   5   4   3   2

Dedication

This book is respectfully dedicated to the women and men of NASA, who, for almost forty years, have made the dream a reality. Inspired by a common vision, they have been almost invincible in their conquest of the universe. Their success has, however, come at a high cost. Working long hours, often in difficult sur-roundings and always under the scrutiny of the public and the world press, many have made enormous personal sacrifices, a few their very lives. Some of their names you know, most you don't, but all, like their feats, are legend. These are the people who will continue to push back the frontiers of space.

And to my dear wife who has made all my dreams come true.

Series editors
Edwin F. Strother, Ph.D.
Donald M. Waltz

Contents

About the Author

Frank Hoban has spent more than forty years in the aerospace industry. A native of eastern Pennsylvania, he joined the U.S. Army serving as a paratrooper with the 82nd and 11th Airborne Divisions. In 1960 he graduated from Parks College at St. Louis University and joined the Peninsula Airport Commission as its Assistant Executive Director. In 1963 he left airport management for the challenge of NASA. In 1970 he was selected to serve on the staff of the NASA Administrator as an Assistant Executive Secretary, first to Dr. Wernher von Braun and later George Low, the NASA Deputy Administrator. In 1973 he joined the newly formed Low Cost Systems Office as the Director of Business Practices. He earned a NASA Exceptional Service Medal for his low cost work. In 1976 he was selected as a Presidential Exchange Executive and spent the next year with the General Motors Corporation in Detroit, Michigan. Returning to NASA, Hoban was assigned to an agencywide task force to implement the historic Civil Service Reform Act. NASA won a major award for its work. As the activities of the task force were winding down, Hoban was asked to join the President's Commission on the Accident at Three Mile Island as its Director of Administration. The workings of the Commission were praised by the White House, the Office of Management and Budget, the *Washington Post* and *The New York Times*. In 1982 Hoban joined the Space Station Task Force as its fourth employee. In 1986, after the *Challenger* accident, Hoban was tasked to lead an agencywide program to train the NASA workforce engaged in managing programs and projects. This effort, known as the Program and Project Management Training and Development Initiative, was very successful and earned him NASA's Exceptional Achievement Medal. He retired in 1994 as a Senior Fellow at George Mason University in The Institute of Public Policy.

Hoban earned an M.A. from George Washington University, an M.B.A. from the College of William and Mary, and attended summer study programs at the Massachusetts Institute of Technology and the University of Utah. He has written extensively about project management and also teaches at the European Management Center in Brussels, Belgium.

Preface

Change is the nature of things for virtually every public and private institution. Government bureaus, corporations and even universities are being reinvented, re-engineered, downsized or merged at an accelerated rate difficult for even their participants to comprehend. Their workers are stunned, invigorated, numbed or bemused as they either recoil from or embrace the change.

This book is about change, change needed for survival. This book is about institutional culture, and it's also about cost control, the mechanism most often used to bring about change. It's about a successful manager and a visionary leader, George M. Low, who was faced with a choice worthy of a Greek tragedy—radically changing the organization he helped create in order to save it. NASA was an organization that less than a year earlier had accomplished a feat so brilliant as to gain the admiration of the world. Friend and foe alike acclaimed the achievements of NASA's lunar landings. The success of Apollo was so complete, so thrilling and uplifting, that NASA received unprecedented recognition and respect. But this book is also the story of an organization so locked into its past as to completely ignore its future.

Only a few of its leaders in the 1970's recognized NASA as an institution filled with uncertainties, and fewer still could imagine the decline of the world's premier space management organization. Now, only a handful can look back to realize the low cost efforts that were tried, those that were not tried, those that worked or failed, and those successes and failures that are still with us today. We either learn the best practices of the past or we try them over and over again.

Change is a dynamic that redirects our plans, challenges our achievements, frustrates our expectations and thwarts our dreams. Change can come instantly, as a storm, or slowly and deliberately as revolution. We have even personified change to the extent that one measure of an individual's character is the ability to "handle" change. We understand the power of change by the exertion needed to manage it. We know that when we ignore change, it still persists; when we force change, the expected results do not always occur; when we deny change, it can paralyze us. But when we accept change, we discover a clearer way of thinking that offers growth and renewal but at times in the form of painful and often difficult choices. The difference between today's headlines and what you are about to read is that this struggle took place not last year or last month but two decades ago. Much of that experience is applicable to today's rapidly changing management environment.

At the beginning of the 1970's NASA was about to complete the Apollo Program, but it had no new national goal to take its place. Without a new headline-grabbing program of national importance, NASA would become just one more Federal agency; in fact, NASA had developed an institutional bureaucracy that rivaled many old line Federal bureaus. The studies that were conducted on NASA's post-Apollo future all stressed a similar theme—NASA had to perform at a much lower cost. Never again would a NASA program be given the resources bestowed on the Apollo Program.

NASA had sold the Space Shuttle as an Apollo follow-on, but this may have helped to reelect a sitting president rather than to serve as a national goal. The Shuttle was an important aerospace challenge, but was nowhere near the scope of Apollo. In its initial configuration, the Shuttle promised NASA and all space users a new, expedient opportunity—ready access to space at a reasonable cost. The Shuttle was supposed to be the key to a new era of low cost, productive use of space.

George Low, NASA's brilliant Deputy Administrator, recognized both the perils and opportunities of the post-Apollo period. He alone among NASA top management was determined to move NASA away from its previous philosophy of success at any cost. Low envisioned using the Shuttle as a way of converting NASA to a low cost supplier of space services. By providing access to and operations in space at a reasonable cost, an entirely new customer base would be able to conduct unprecedented space-based commercial ventures. Nothing would help NASA more in its struggle to survive than the development of new industries, as it had already shown in the success of the satellite communications industry.

This book follows Low's progress from the initial recognition in 1972 of the need to be much more cost effective, to NASA's attempt to find a vision for the "new" space agency. It covers the rise, rejection and fall of the Low Cost Systems Office in NASA, an idea whose time had not yet come.

This book attempts to explain why the low cost program fizzled, and why the same or similar cost issues still plague NASA, government and industry a generation later.

The book includes specific recommendations for low cost programs in NASA, government, and industry. It concludes with a look at the ultimate in cost reduction and an examination of the NASA of today and its relevance to the civil space program of the next decade.

Acknowledgments

I have had much help and support in the preparation of this book. In addition to those named in the text and on the cover, my thanks go to the following people:

First of all, to Kingley Haynes, Roger Stough and Arthur Melmed of The Institute of Public Policy at George Mason University for their kind encouragement and ample support.

To my many colleagues at NASA who contributed generously to this book with their recollections, counsel, corrections and interpretations. Chief among them are John Sheahan, Tony Schoenfelder, Bryant Cramer, Richard Daniels, John Rellar and the entire History Office, especially Roger Launius, Lee Saegessar and Nadine Andreassen. Also, with their rich NASA experience, thanks to Shirley Malloy, Ronald Muller, Norman Gerstein, William Cunningham, Joseph Purcell, Raymond Tatum, Robert Lindley, John Hodge and Lawrence Vogel.

To Debbie Duarte, James Muncy, Joseph Keogh, Michael Gillies, Robert Burns, Paul Muraco, Penny Trusty, Gerald Winchel, Gary Clemmons, Arthur Wisely, Harold Wood, Morgan Jones and T. Martin Scally who read portions of the multiple drafts and offered valuable suggestions.

To Sarah Lowe and Tracey Martin who provided logistical and administrative support.

To my editor, Mary Roberts, and all the professional staff at Krieger Publishing Company.

To all those who helped with this massive project whose names are not included above. You know who you are, and I am grateful for your encouragement, help, advice and support.

And to Jeffrey Michaels, my special thanks to a most talented young man.

Introduction

Today's National Aeronautics and Space Administration is an aging artifact of the Cold War. From its inception in 1958 until the Apollo challenge, NASA was busy sorting out its new organization and managing a few diverse programs, such as the Saturn launch vehicle and Project Mercury, but it had no firm, future direction.

The assembly of this new agency, from as disparate a collection of organizations and cultures as ever seen, was right out of Mary Shelley's *Frankenstein*. The world renowned staff from NACA, the National Advisory Committee for Aeronautics, a tight knit collection of very conservative engineers, formed its core. From the Army's Ballistic Missile Laboratory in Huntsville, Ala., came the Germans, the only team in the world with extensive operational missile experience; the Naval Research Laboratory folks, along with the researchers and managers from JPL, the Jet Propulsion Laboratory in Pasadena, also filled the ranks. However, when President Kennedy gave NASA the Apollo challenge, these and many others from industry, the DoD and academia came together and formed an almost perfect management team, one capable of accomplishing the unbelievable on time and within advertised cost, a feat NASA has since had difficulty repeating.

With the Congressional approval of Apollo, NASA had a unifying program of an unprecedented magnitude, a program that quickly became NASA's raison d'etre. NASA was Apollo and Apollo was NASA, and for almost a decade NASA existed to manage the lunar landing program and, more importantly, to contribute to the decline of the Soviet empire.

To manage the new, focused space agency, President Kennedy needed an extraordinary leader he could depend upon. James Webb, NASA's Administrator from 1961 to 1968, had a long and distinguished career in government before he accepted the leadership of NASA. At age 40 he reported directly to President Truman as Head of the Bureau of the Budget. Truman was so pleased with Webb's performance that he made him number two at the Department of State. Webb learned how to get things done in the Congressional morass. He gladly accepted Kennedy's Apollo challenge to demonstrate to the world, in a program open to full scrutiny, the preeminence of American science, engineering and technology and, by inference, our political system over that of the Soviet Union. He also knew that Congressional support for even such a dramatic program could be fleeting.

To ensure the longevity Apollo needed, approximately 10 years, Webb devised a new government management model. It would be different from the model used by DoD in their successful management of their Polaris and Minuteman programs. Both of these programs, although not of the scope and public scrutiny of Apollo, were large, complex systems that presented distinct management challenges. In the DoD model, the government relies heavily on industry to manage the design and development of its systems. Industry is the dominant player. The government provides the requirements, establishes a small project office, and, in general, allows industry to get on with the job.

Webb, on the other hand, elected to use a management process in which NASA and industry engaged in a "partnership." In this model, the government, through its agent NASA, was the dominant partner. NASA not only told the industry what it wanted, but directed how industry should do it. NASA would actively participate in the day-to-day project management, including problem solving right down to the subcontractor level. This model, amounting to near nationalization of the aerospace industry, required new NASA Centers of expertise and large numbers of highly skilled civil servants and support contractors. When Webb designed his management scheme, he gave the powerful chairmen of the House and Senate Appropriations and Authorization committees major stakes in NASA and Apollo by establishing a string of facilities across the South in what came to be known as NASA's "southern crescent." These facilities gave the southern region of the United States an entree to the world of high technology and an escape from the burden of an agrarian economy. In return, Webb had years of favorable budgets and the stability Apollo needed.

NASA's aeronautics and planetary programs, as spectacular and successful as they were at times, simply did not command the public's attention like human missions. When the last astronauts departed the lunar surface and splashed down safely in the Pacific Ocean, Apollo was over, and in the eyes of many, so was NASA.

NASA survived Apollo. However, the result was a government-dominated aerospace industry with an inbred culture where cost was not an important criterion for success—an industry not accustomed to doing programs in a cost effective manner. This new aerospace industry was nursed by government contracts that ensured all their legitimate costs would be paid, along with a handsome fee if they just gave their "best effort." With the equivalent of a key to the Treasury vault in their pockets, industry had absolutely no incentive to worry about cost. In fact, it would have been counter-productive to their bottom line. Industry also enjoyed the luxury of blaming at least some portion of any failure on the government; after all, they told us to do it that way. Quasi nationalization had its rewards.

The government side was not much better off. Post-Apollo NASA lost its raison d'etre. In 1972 it needed new, large, significant space ventures to remain a viable and important agency of the national government, but NASA no longer had strong advocates in the White House or the Congress willing to give the Agency the needed political support. The strategy that Webb had so carefully devised had collapsed by the completion of Apollo, the decline in public support for space exploration in general set in, and the escalation of the Vietnam war overshadowed just about everything else. The war effort brought many more dollars to the southern region than NASA ever could. It has been speculated that one reason President Nixon approved NASA's Space Shuttle proposal, thereby ensuring the existence of the total institution, was through fear of losing California's electoral votes in the upcoming 1972 presidential election. However, the annual funding the Nixon

Administration was willing to devote to the Shuttle was far below that needed to design and develop a cost effective space transportation system under the Apollo management model. It was no wonder, then, that NASA slipped into a self-preservation mode after Apollo.

The immediate task facing NASA's dynamic and energetic Deputy Administrator George Low in 1972 was the survival of a missionless NASA and the country's civil space program. He was well aware of the anger and dismay of the NASA and contractor work force over the apparent lack of appreciation shown by the White House, Congress and the public for their past accomplishments. And he was painfully aware of the dangers of a fragmenting NASA organization, each Center selling pet projects, sometimes at the expense of the whole agency.

Low believed he had to change NASA to save it from itself. In the interim he tried desperately to keep the institution occupied by involving it in a wide-range of non-space activities. The Jet Propulsion Laboratory supported a variety of DoD initiatives, and other NASA Centers engaged in energy-related research and development programs, such as windmills, insulated housing and more efficient ground transportation. NASA supported President Nixon's 1972 New Technology Initiatives Program, attempting to curry favor with this otherwise disinterested Administration, and in general delved into work with a more terrestrial bent. In many of these endeavors NASA provided elegant, but high priced and sometimes impractical, engineering solutions. However, all of these efforts were too small to have any significant impact on the NASA institution, an institution carefully and skillfully crafted by Webb to plant this nation's flag on the moon. Should NASA have taken the bold step and reorganized and downsized to reflect the reduced missions of the 70's more properly? There is some evidence that, at the urging of the Administration, NASA reluctantly attempted to relinquish control of one Center to another government department, but this was nipped in the bud by a powerful Senate chairman.

So, this was the foundering NASA of the early 1970's. Despite the fact that NASA's hasty, desperate, post-Apollo proposals were virtually laughed out of the OMB and the White House, many in NASA were certain that eventually the Administration and the Congress would come to their senses and again ask NASA to manage a heroic program of vast importance to the nation if not the world. However, the only new major agency-wide program was a downgraded and underfunded Space Transportation System. In the ranks, managers were questioning if NASA really needed all the employees left over from Apollo, as well as the massive organization to manage much less future work.

Low believed if NASA were to survive, it would have to change radically and such change would have cost as the focal point. He began with the 70's version of today's faster-better-cheaper: the Low Cost Systems Effort. How the effort was started, progressed and ultimately failed is the story that follows. It may be set in the 1970's, but most of the problems plaguing civil space endeavors are still with us today.

In addition, most of the discoveries and solutions found then, in the mid-1970's, are just as valid today as they were during the Low Cost Systems Effort. It is important, therefore, to look back and see what went wrong with NASA's pioneer efforts to identify and control cost. The astute reader will find that the recommendations made and the lessons learned in that earlier effort are worthwhile palliatives to the wrenching dislocations and damage done by heavy handed efforts to downsize, reinvent and reengineer organizations.

It may be discovered that the Low Cost Systems Effort did not really fail after all. It was just that the effort was never really implemented, not really tried. Had it been, NASA could have been a model of the organization of the future rather than a Cold War relic.

Chapter 1

An Agency in Search of a Mission

By the early 1970's, the National Aeronautics and Space Administration was looking back at the Apollo moon landings as "the greatest engineering feat of all time" and free-falling into an uncertain future. After Apollo 11 hardly anyone, even in Congress or the Nixon Administration, paid much attention to subsequent lunar landings, and few seemed to notice that Apollo missions 18, 19 and 20 were quietly scrubbed—due as much to lack of interest as to a dwindling NASA budget. NASA was perceived as "mission accomplished" with nowhere else to go. Despite that perception, NASA persistently attempted to shape its future, an effort that continues to this day.

Planning for the post-Apollo years began well before Neil Armstrong and Buzz Aldrin landed on the moon's surface on July 20, 1969. Five years earlier in January 1964, President Lyndon Johnson asked NASA Administrator James Webb to begin planning the next major mission after Apollo. His concerns were twofold: that a massive layoff of scientists and engineers would occur if NASA had no post-Apollo mission, and that there seemed to be no purpose or mission for the proposed multibillion-dollar NERVA program, a nuclear powered rocket engine. Webb was given a deadline of September to present a plan to the President.[1]

However, it was not until January 1965 that Webb responded with options, possibilities and a series of objections. Webb thought a specific, long-range, post-Apollo program would be an easy target for space detractors and government budgeteers. In addition, a second big space program might detract from the burgeoning Apollo Program, diverting the energies of Apollo managers and possibly impairing the success of Apollo. Instead of unveiling a long-range plan for civilian space efforts, Webb advised Johnson to hold off until later.

Meanwhile, the Manned Orbital Laboratory (MOL) project was begun by an organization more sure of itself and the future, the U.S. Air Force. The FY1967 Federal budget, for the first time, called for a decline in NASA funding.

For FY1968, Webb wanted $6 billion for NASA. He wanted to restore the Agency's budget cuts with preliminary funding for the NERVA study and an Apollo Applications Program (AAP) to make use of leftover Apollo hardware in the 1970s. Both programs were included in the President's budget proposal and were sent to Congress in January 1967.

By the summer of 1967, everything had changed. Opposition to the war in Vietnam had increased dramatically, and President Johnson sought a very unpopular tax increase. The War on Poverty focused the American public's interest on immediate problems on Earth rather than promises in space. The Apollo capsule fire and deaths of three astronauts in January 1967 diminished dreams of space adventures; instead of restoration NASA's budget suffered even more cuts. President Johnson decided not to seek reelection. James Webb resigned in September 1968.

In a speech at Harvard University in 1968, Webb acknowledged that the "national decision" for post-Apollo planning began to fail in 1964 and continued incrementally, year by year. In other words, the decline of NASA began five years before the greatest engineering triumph of all time, the Apollo moon landing.

At about the same time, on another front, the President's Science Advisory Committee (PSAC) formed a prestigious group of Space Science and Space Technology panels to examine America's next step in space exploration. The panels' 1967 plan focused on the moon, calling for one and two manned expeditions per year after Apollo, supplemented by unmanned science missions to the lunar poles, and manned explorations of the polar regions from 1975 to 1980.[2]

Even more ambitious was a post-Apollo plan submitted by NASA to the newly formed Space Task Group (STG) in 1969. NASA called for then-massive amounts of funding ($4 billion to $8 billion in peak annual costs) for a permanent manned space station in low-Earth orbit by 1975 to accommodate 50 to 100 scientists, as well as a "new and truly low-cost space transportation system" (including the Space Shuttle), a lunar orbiting base in 1976, a lunar surface base in 1978 and a manned expedition to Mars in 1981. On the way back from Mars, the manned spacecraft would conduct a flyby of Venus, scheduled for 1982.[3]

Surely, the nation—if not the world—would be radically different had these bold plans been implemented. Why they were not pursued may have had much to do with the economic climate, but the NASA report anticipated objections from the Nixon Administration budget watchers by stressing "low cost access" to space after Apollo.

> The key to carrying out this step in an effective and efficient manner will be commonality—the use of a few major systems for a wide variety of missions; reusability—the use of the same system over a long period of time for a number of missions; and economy—the reduction in the number of "throw-away" elements in any mission, the reduction in the number of new developments required, the development of new program principles that capitalize on such capabilities as man-tending of space facilities, and the commitment to simplification of space hardware.[4]

It did not happen. While NASA may have anticipated a low cost systems effort in the 1970s, its Cold War rhetoric seems to have fallen on deaf ears. "The Soviet space challenge is not trivial," they warned, inadvertently suggesting that NASA may eventually become a Cold War relic.

Another reason for the post-Apollo decline of NASA could be called "managerial calcification" or rigidity. Former NASA Administrator James E. Webb warned as early as 1962, in a speech to the American Society for Public Administration:

As organizations grow to maturity, they have a tendency to become much less flexible in their managerial thinking. To prevent this from happening is a major challenge to general management in technologically based organizations such as NASA, and will present one of the continuing challenges in the administration of the civilian space program.[5]

Space historian Howard E. McCurdy agrees. In trying to show the cultural change as NASA entered its second decade of existence, he said:

As NASA matured, it lost much of the administrative flexibility that characterized the early years. Bureaucratic procedures proliferated . . . Officials became more concerned with the survival of the organization. They adopted management schemes that sacrificed institutional flexibility in order to ensure institutional survival.[6]

Nevertheless, the press and the public, no doubt proud of the Cold War achievement of beating the Soviets to the moon, seemed to turn their attention to other things, most notably to ending the war in Vietnam and getting control of a runaway economy. Anti-war protesters were in the streets and taking over college campuses, amid talk of wage freezes and price controls. America's mood at the end of a turbulent yet triumphant decade was neatly summed up in the nation's leading pop-political picture magazine. The December 26, 1969, *Life* magazine was a special issue devoted to the decade of the 1960's. It covered the troubles of the times—the war in Vietnam, the riots, the assassinations. But it also singled out the one event that perhaps made the rest of the turmoil bearable. As *Life* put it, "The decade ended with an adventure so fantastic as almost to overshadow and redeem all turmoil that had scarred it." That adventure, of course, was the Apollo 11 lunar landing. But for NASA, the Apollo 11 mission was a high point of a different kind. It marked the beginning of a precipitous decline in support from the administration, the Congress, and the public.

While many people today may look back and think that enthusiastic supporters of the Apollo Program declined in number during the 1970's, Gallup Poll data from the American Institute of Public Opinion suggests a more important fact—that opponents of space spending increased. Herbert E. Krugman, manager of public opinion research at the General Electric Company, observed that Americans seemed to be turning away from international concerns like the war in Vietnam and the "space race" with the Soviets. "However," he notes, the data suggests that it was the "successful" landings themselves which also spurred the increase in opposition. The public had been exposed to at least a decade of intensive media attention to "getting to the moon." It had been a topic of curious speculation for

Apollo 17's 17 Minutes of Fame

Air & Space magazine for November 1992 recalled that no one seemed to complain when networks cut back or dropped coverage of Apollo launches. It seems far more people were interested in the outcome of "Medical Center" when CBS cut away for the nighttime launch of Apollo 17.

centuries. Once the first landing occurred that interest was forever terminated. In terms of public curiosity, there was little point in the subsequent moon landings. There was "nothing more to be done," and it was a "waste," said the opponents. In this sense public support for the Apollo program had been designed to self-destruct on the initial achievement of the program's major objective.[7]

Some sense of the decline of interest and support can be found in an opinion piece written at the completion of the program by the acclaimed Chicago-based, nationally syndicated columnist, William Hines, which appeared in several newspapers of December 12, 1972, titled "End of a Crazy Business." Hines began: "And now, thank God, the whole crazy business is over." He even counted the days: "For 11 years, 6 months and 24 days—John F. Kennedy's proclamation of men on the moon in the '60s to the splashdown of Apollo 17- the United States was on a moon bender that cost $25 billion in cash and an incalculable amount of emotional energy.

"It started on May 25,1961, born of Vostok 1 and the Bay of Pigs and the same kind of mindless machismo that makes a virtue out of the Katzenjammers. It ended on December 19, 1972, with neither a bang nor a whimper; actually, more of a ho-hum. The final Apollo flight, in many ways the most interesting and significant, was almost ignored; at least in comparison with earlier trips to the moon."

Hines summed it up: "In 3 1/2 years of studying lunar rocks they have refined their theories a bit about the origin of the moon—a neater fix, so to speak, on the number of angels who can dance on the point of a pin." He added: "Project Apollo has not made the world a better place to live in, or life more worth the living." So, he concluded: "Thank God the crazy Apollo business is over."

Even before Hines' vitriolic piece appeared, concerns about the space program, its costs and its very existence had appeared in national and international newspapers and magazines. NASA took a constant beating from the internationally esteemed *New Scientist* magazine. "Let us stress once again, and emphatically, that the true scientific value of the Apollo project is virtually nil," they said at the end of 1968. "The American taxpayer would get infinitely more for his investment if it were put to promising lines of Earth-based research." Another analytical article appeared in *Innovation* magazine by senior editor Englebert Kirchner, a longtime NASA watcher, titled "Sorry Virginia, There Is No Space Program." Kirchner attempted to answer the question of what was happening with the U.S. space program by questioning the existence of such a program. He wrote, "Most of the 30,000 people in NASA spending their time are simply doing their jobs. But in the upper echelons, there are several hundred people, including a high proportion of managerial brains of the first order, whose consuming interest, day in and day out, is to figure out what is best for NASA and to try and get it. And these hundreds can always call on the thousands below them for ideas and calculations and lines of argument."

Kirchner pointed to the importance of the NASA *culture* instead of the NASA *mission*: "Even to most of the people in Washington who live and breathe the institutional politics of government, the fate of NASA is a sometime thing. To NASA, it is everything. Which brings me to the point of these sociological excursions (not to be attempted where the cocktails flow): It doesn't really matter whether a space 'program' exists or not, a blueprint drawn to some more or less objective standard of what we should do in space. What we will do primarily depends, in any case, on how NASA fares in its struggle to preserve its *institutional existence*."

It was becoming clear that the Cold War rationale, which served as impetus for the Apollo effort, was no longer pertinent, for even the Soviets seemed tired of space. On April 29, 1971, Joseph Kraft, in a nationally syndicated column titled "Space Shots Bore Russians," discussed the apparent apathy of the Russians citizenry to their highly successful space program: "But all of this is lost on the Soviet population. Public interest in space shots here has been remarkably slight. Many educated Russians seemed not even to know that a new effort was under way. Even after the news had spread all around town, ordinary Russians were not moved to turn on the television sets in the hotel lobbies to get late reports. A Soviet economist to whom I talked about space also acknowledged that public interest has waned. 'People are used to space now,' he said. 'It is considered ordinary and everyday. Nobody gets very enthusiastic about what happens up there.' "

Loss of public interest may have been of little consequence in a controlled economy like the USSR, but in a democracy, where competing interests fight for public dollars, public support was vital for an Agency like NASA. Without it, NASA would decline.

NASA's Response: The Think Group

Some of NASA's leaders had long been concerned about the coming of such a decline. Just a few days after the Kraft piece, in May of 1971, NASA Deputy Administrator George M. Low put together a group of senior NASA officials to address this problem and to explore what the Agency could do in the remaining three Apollo missions—15,16 and 17—to rekindle the interest of the American public. This informal collection of some of the best minds in NASA was nick-named the "Think Group." It consisted of George Low; Dr. Wernher von Braun, Chief of NASA Planning; Dr. Robert Jastrow, Director of the Goddard Institute for Space Studies; Dr. Homer Newell, the Associate Administrator and NASA's Chief Scientist; Edward Cortright, long time Low colleague and Langley Research Center Director; Richard McCurdy, Associate Administrator for Organization and Management and former CEO of Shell Oil Co.; the program associate administrators and others as invited. The Group met after hours, at informal get-togethers hosted by individual members.

Low's timing on the formation of the Think Group had much to do with the landing site selected for the Apollo 15 mission. In contrast to the stark, flat, non-dimensional sites of previous missions, this mission would explore the Hadley-Apennine region, a mountainous area with the added attraction of a lunar Grand Canyon, Hadley Rille. In addition to the enticing topography, Apollo 15 featured the first color TV broadcast from the lunar surface and the introduction of the lunar rover, an electric powered four-wheel-drive vehicle. The rover would enable

Those Opposed
The more things change, the more they remain the same.
 In a column labeled "Database," U.S. News and World Report (June 7, 1993) reported that only 39% of Americans in 1970 said it was worth the cost to land a citizen on the Moon. Those who said it was not worth the cost: 56%.
 In 1991 the percentage of those who said the percentage of money for space activities would be better spent terrestrially on medicine and schools was identical: 56%.

A Post-Hubble NASA

Time magazine for Dec. 20, 1993, lashed out against NASA's "badly checkered history" and even wondered aloud if the U.S. really needs a space program at all. In a final jab, Time questions whether NASA should be the lead agency or not in a space program after the spotted Hubble mission.

The Lunar Rover on Hadley Rille. *Photo courtesy of NASA*

the astronauts to roam and explore well beyond the confines of the landing site. Conditions would never be better for a rousing TV spectacular designed to return middle American viewers to their sets and the U.S. space program.

The group considered a wide range of options. They spent a great deal of time debating the value and appeal of watching an astronaut pushing a boulder into Hadley Rille. The slow motion fall of a huge rock into a deep canyon at one-sixth gravity would prove incredible, but the scenario was dropped when the danger was pointed out of the astronaut tumbling in behind the boulder. Several attempts at viewer stimulation were made during the remaining missions, such as the feather and hammer drop and racing the Lunar Rover, but these were pale versions of what the group had hoped for. Even with the help of a former TV network executive, the Think Group failed to create a TV extravaganza.

The group then attempted to identify a theme that they could use to build a vision for the future. Why was mankind, especially the U.S. and specifically NASA, in space? Why were we going to the moon? Why were we spending increasingly scarce tax dollars on exploring a seemingly empty cosmos? The answer to these questions would provide the vision for which the group was searching.

On May 20, 1971, the newly appointed NASA Administrator, Dr. James Fletcher, spoke to the Aerospace Industries Association (AIA) in Williamsburg, Virginia, on just those questions. Fletcher's speech, "Benefits of the Space Program," had been worked over extensively by the entire Headquarters speechwriting staff, along with a number of other senior NASA officials. They thought the address they had painstakingly put together was a real winner. In this speech, Fletcher recognized that NASA and industry had both been wrestling with the problem of winning and keeping the public's support, and that this struggle had been going on for many years. Fletcher asked his audience for new ideas, for he intended during his administration of NASA to make this issue the highest of priorities. He cautioned the audience not to search for a single overriding justification for the exploration of space because there was none; space benefits could take many different forms, both tangible and intangible. Fletcher then gave five reasons he believed the nation should continue its substantial investment in space:

1. Some satellites, like weather and communications, pay for themselves by doing useful work. "We should invent and fly more self-supporting satellites."

2. "We are seeking and getting valuable scientific knowledge from space we could not get in any other way—valuable new knowledge about the Earth and its atmosphere, the sun and the planets, and the universe. And about human life itself."

3. "Our national security is at stake in space." It would not be safe for the United States, with its great responsibilities for world peace, to lag behind any other country in space technology. "This is an axiom we did not quibble about in the 60s and should not quibble about in the 1970s." A strong space program offers many new opportunities for significant international cooperation, and it promotes the cause of world peace.

4. The space program has proven to be an excellent hotbed for forcing new technology, which in turn raises our national productivity and prosperity, and increases our ability to solve pressing social problems of today's urban society. "This is a message we really need to get across."

5. Looking beyond the material benefits, Fletcher concluded, space exploration provides inspiration for all people. "We cannot measure this, we may not be fully aware of it, but I think we are inspired, and our children are inspired, to be living in an age when men first moved out into space and began the exploration of our solar system. I think we would be ashamed of ourselves, as a society, if we withdrew from space exploration now after such an auspicious beginning."

Bullish on the Future of Space

Today, the United States has a space program that plays only a minor role in the public agenda and has much of its appropriations justified on the basis of government-related jobs that it provides. In substance, it has lost its commanding constituency of the 1960s.

Space News— September 13-19, 1993.

These and subsequent selections are reprinted courtesy of Space News. Copyright by Army Times Publishing Company, Springfield, Virginia.

Forcing new technology was the main theme of the address. Fletcher touched on technology throughout and returned heavily to it in the closing minutes when he reminded the audience that the key to success of a modern industrial state was productivity, and the best way to increase productivity was to advance the nation's technology. NASA and the space program, in his opinion, were the best vehicles for that. Fletcher then stepped into the technology transfer trap that NASA had been building for itself for years when he said: "I think we have to look for the benefits of new technology at the other end of the progress, by working backward. Let us look at the areas of great technological growth during the last 10 years—computers, communications, medical science, lasers, automatic controls of all kinds, sensors of all kinds, new standards of quality control, and so on. Let us look at each of these fields and ask the question: Have NASA requirements and NASA procurements played a significant role in the phenomenal technological progress being made in this field? The only honest answer has to be, 'You bet they have.'"

People in NASA firmly believed this to be true. In numerous instances, NASA was convinced it had pushed technology to new heights in order to meet the demanding requirements of the space program—advances that would not have occurred without a space program. The Think Group decided to follow up on the technology theme and asked for industry backing in the preparation of materials to use on television during subsequent missions.

The Think Group was especially anxious to get this message to "middle America." There were two particularly attractive technology application items to brag about—highway and runway grooving, and the performance of the weather satellite system during Hurricane Camille, which NASA was certain had saved 50,000 lives. Indeed, on June 23, 1971, NASA Administrator James C. Fletcher testified before the Senate Appropriations Subcommittee on HUD, Space and Science: "When 50,000 lives were saved as a result of satellite advance warning at the time of Hurricane Camille, the space program probably paid for itself . . ." The Think Group asked the director of NASA's Technology Utilization Office to survey industry as to what technology advancements they would credit to NASA.

Not a single major company agreed with NASA's assumption that the space program had been a major technology driver. If that were not enough, they were told that runway grooving was developed by the English shortly after World War II. The group was also told that if all the inhabitants of southern Louisiana had been made to stand on the shore of the Gulf of Mexico when Hurricane Camille hit, it could not have killed anywhere close to the number of people NASA claimed to have saved. To their credit, the Think Group did not shoot the messenger, but feelings of disbelief, dismay and betrayal swept over them. It seemed all that was left to sell were Teflon pans and Velcro straps, although NASA's role in the development of these was even disputed.

It was not just industry who had doubts about NASA's role as a prime technology driver. Consider this exchange at a Senate Appropriations Committee hearing of March 13, 1974:

Dec. 21, 1992

Daniel S. Goldin, Memorandum to the Officials-in-Charge of Headquarters Offices Directors, NASA Field Installations Director, Jet Propulsion Laboratories –Washington, DC

 "NASA has the reputation of being the leader in technology transfer, but this position has eroded. Our attitude that the transfer of our valuable technology will 'just happen' is no longer acceptable; it must be proactively sought and given the highest priority."

8

Senator Metzenbaum: . . . I must confess I am not at all certain that as the economists say, in 15 years we will bring back a number of dollars. It is pretty hard to get a handle on this. When you talk about some of these things, I wonder if instead of spending $40 billion, if we had spent $300 million or $200 million, we wouldn't have been able to produce all of the same kind of technological developments.

NASA's response by the head of the Office of Industry Affairs and Technology Utilization, Ed Gray, was a little vague:

Mr. Gray: Well, Senator, I think it is quite evident that this country is in a competitive race for its position in the world in the economic sense, and that there are a great many economists who have been viewing with a great deal of concern that we are losing our position in the world markets and that there is a great need to increase the R&D in this country; and I think from everything we have seen so far, that the important thing is to do R&D across a broad field of technology, and I think that is one area that the space program has proven to be invaluable in, in that the type of problems that were involved in creating R&D is sensitive to all these areas.

Senator Metzenbaum: I agree with you that we are in a competitive position. We are also in a competitive position in this country for the R&D dollars, whether you use the R&D dollars for the development of new energy sources or for the development of new research in the health field, or in a host of other areas. It seems to me that the space program is in competition with the technological—in its technology in other areas. Whether money can be more effectively used to do the very things that you just mentioned, or whether it can be used better in some other way.

The question that NASA failed to answer that day was about the advanced heart pacemaker. Would it not have been faster and cheaper to develop with a focused program directed solely at the pacemaker? Following is the end of that senatorial exchange:

Senator Abourezk: I think Senator Metzenbaum, if I understand it correctly, is trying to say that the space Agency tries to justify its manned space program by talking about all these spinoffs that they get, where these particular technological benefits, advances could be done a great deal cheaper without the manned space program. And as I understand what you are saying, you keep talking about that this is great research and we have to keep up ahead.

Well, of course, we do: but we don't have to keep a manned space program going necessarily. I think the unmanned space program is fine, and I think we got a lot of spinoffs; but we do it relatively, in a less expensive

An article in the March 25, 1994, Washington Post, "Budget Cuts May Squeeze Man Out of Space," addresses the criticism that NASA's advances do not contribute to the competitiveness of U.S. industry, according to the Congressional Budget Office.

manner with relation to the manned space program and I agree with Senator Metzenbaum in that regard.

Senator Metzenbaum: I don't criticize the NASA program for talking about its fallout, I actually commend them for that. I think what I am trying to say is that the value of the fallout is quite small as compared to what it would cost us to get the same kind of research and development if it weren't in the space program.

Still No Consensus

A Space News headline from summer 1993 reads: "Group Fails to Reach Agreement on NASA's Purpose."
  By 1993, NASA was still looking for its vision. A year after the NASA Administrator ordered a strategic planning group to come up with a vision for NASA, the Agency was still undecided on where it was going.
  Alternative and competing visions included preservation of the environment, stimulation of economic growth, support of educational excellence, and promotion of peaceful exploration of the universe.

After the shock of the industry response wore off, George Low asked Dick McCurdy to take a hard look at some of NASA's technology advancement claims. McCurdy gave a verbal report to one of the Administrator's staff who wrote a note to Low: "To quote Mr. McCurdy, 'If the Federal Trade Commission investigated all our claims, we would be in serious trouble with the government. I guess in the case of [Hurricane] Camille, Tom Paine would be in jail.' " (Paine was the previous NASA Administrator and the one to first make the unfounded claim.)

But the group did not quit. If they could not sell technological innovation, they would simply find another theme to shape into a vision. They felt a need to answer the question "Why are we in space?" Each member of the group prepared a short paper addressing the question. The completed papers stressed:

- Development of practical applications

- Substantial contributions to the advancement of science and human knowledge

- Survival of the human race

- Opening of new frontiers

- Uplifting of the human spirit

- National security

- Major advances in technology, including space manufacturing (for those not yet willing to give up the issue)

- NASA's ability to manage complexity

von Braun was not at all convinced of the last point, and he suggested to the group that it had been easy to go to the moon. After all, he said, there were no people between here and there. He was not convinced that NASA's management track record could withstand the challenges of complex social programs such as housing and education, where the emphasis would be on people instead of hardware. He also reminded the group that the most successful marketing ploys of the past might still be useful, namely: "the Russians are coming" and the appeal to aged Members of Congress that space discoveries could extend their lives (if not their terms of office.)

A draft consolidated paper was prepared covering four topics: national security, survival, human nature and possible benefits from space exploration. In the end, it was decided that all the materials developed by the group would be used in some form, and included a backup book for the Apollo 15 mission. However, Low called no further meetings, and the Think Group faded away.

Footnotes

[1] W. Henry Lambright, Presidential Management of Science and Technology: The Johnson Presidency. Austin: University of Texas Press, 1985, pp. 142-150.

[2] President's Science Advisory Committee, "The Space Program in the Post-Apollo Period." Washington, D.C., 1967.

[3] NASA's "America's Next Decades in Space: A Report for the Space Task Group," and "Goals and Objectives for America's Next Decades in Space," both dated September 1969.

[4] NASA, "America's Next Decades in Space," September 1969, p. 6.

[5] James E. Webb, "Administration and the Conquest of Space," April 13, 1962.

[6] Howard E. McCurdy, Inside NASA: High Technology and Organizational Change. Baltimore: Johns Hopkins University Press, 1993.

[7] Herbert E. Krugman, "Public Attitudes Toward the Apollo Space Program, 1965-1975" in Journal of Communication, Vol. 27, Autumn 1977, p. 93.

Chapter 2

What We Really Need Is Low Cost Access to Space

By early 1972, NASA's importance to the Nixon Administration continued to decline, Congress continued to be absorbed with other pressing agenda items, especially the war and the economy, and the public's interest in space continued to plummet. George Low, more than any other NASA official of this time, was aware that a continuation of this trend could threaten the very existence of the NASA he helped build. He believed NASA must change the way it did business, and this change must make space readily accessible, much less expensive, and attractive to a broad variety of new users. A start in this direction would be to explore what was necessary to convert NASA into a low-cost provider of space services.

Cost had not been the primary concern during the Apollo years. Schedule and performance were far more important, to satisfy President Kennedy's mandate and beat the Soviets to the moon "within the decade." However, with the race to the moon won, all of a sudden NASA, like other Federal agencies, had to compete for and justify its portion of the tightened Federal budget. Low was ready to accept this challenge.

George Wilhelm Low was born on June 10, 1926, on the family farm near Vienna, Austria. In 1934 his father died, and four years later, during the rise of the Hitler era, the family fled to Switzerland, England and then to Forest Hills, New York. He studied aeronautical engineering at Rensselaer Polytechnic Institute (RPI), a school he returned to periodically for the rest of his life. World War II interrupted his studies at RPI. Low served in the U.S. Army from 1944 to 1946 as a topographic draftsman, returning to RPI for his bachelor in aeronautical engineering degree. He earned his pilot's license then and became a naturalized citizen in 1945, changing his name to George Michael Low. His first job was with Convair in Fort Worth, Texas. In 1948 he returned to RPI to pursue a Masters degree and married Mary Ruth McNamara. They had five children, including a future astronaut, David.

In 1950, Low joined the National Advisory Committee for Aeronautics (NACA) as an aeronautical research scientist at the Lewis Flight Propulsion Laboratory in Cleveland, Ohio. Low jokingly recalled that Langley Research Center had turned him down and that Lewis was his second choice for entering NACA. Following Sputnik in the summer and autumn of 1958, he supported the

Small Launchers Slowly Progressing

Low-cost access to space remains the biggest obstacle facing the growth of the light satellite industry, according to industry officials.

"Lighter, Cheaper Payloads Call for Smaller, Low-Cost Vehicles to Space"

Space News—
August 23-29, 1993.

What Does the Future Hold?

The high cost of space launch has been identified in study after study over the past 25 years as being a pressing national need. In the past five years alone, the United States has spent more than $3 billion on a wide variety of ways to lower costs, but has practically nothing to show for it.

Space News— May 16-22, 1994, p. 15.

planning for the new Agency and also the planning for Project Mercury. Soon after the formation of NASA, Low transferred to NASA Headquarters as the chief of Manned Space Flight. He participated in the planning of projects Mercury, Gemini and Apollo, and was instrumental in developing the lunar landing proposal for consideration by President Kennedy. He rose through the Headquarters Manned Space Flight Program ranks until 1963, when he was named Deputy Associate Administrator of the office. In 1964, he accepted the position of the Deputy Director of NASA's Manned Spacecraft Center in Houston, Texas, where he functioned as the Center's general manager. Under his direction as Apollo Spacecraft Manager, eight flights were successfully flown, including Apollo 11. He won the admiration and respect of NASA and industry colleagues during his tenure in this position. He especially counted on the outstanding NASA and industry work force, eager for a new challenge after the spectacular success of Apollo, to fuel his vision.

In December 1969, Low was appointed NASA's Deputy Administrator serving as the Agency's general manager. In the context of the post-Apollo transition planning, Low became convinced of the need for heightened cost awareness and program control. Low's vision was to make space attractive and available to a whole new range of customers—the entrepreneurs, the small science advocates, the low budget payloaders, the venture capitalists and the space pioneers—in fact, anyone in good health was a potential Shuttle payload investigator.

George M. Low.

Photo courtesy of NASA

After all, a joint Department of Defense (DoD)-NASA study team had assured President Nixon that a fully reusable Space Shuttle system could be built in eight years for a mere $5 billion. The DoD-NASA team estimated that after R&D was completed, each Space Shuttle could be built for about $250 million. The per-flight cost would be about $5 million, so the Shuttle's 50,000-pound payload could cost $100 per pound for each launch into low polar orbit.[1]

NASA was still estimating Shuttle development cost at $5.2 billion in June 1972 at a cost per launch of $7.7 million.

Low's resolve to reduce the cost of space transportation, utilization and exploration had taken on a new urgency during a February 11, 1972, meeting with the Office of Management and Budget (OMB). The meeting to discuss the Shuttle decision process concentrated instead on the need for NASA to stay within the agreed constant annual budget plan, even if that meant reducing the size of the Shuttle or using less expensive solid rocket motors. This news was followed by a second jolt when the constant budget plan revealed at the meeting was almost $300 million

14

below the mark NASA thought it would have. In less than a week after the OMB meeting, Low made his first address outside the Agency on the need to reduce cost.

In the speech, "Productivity in the Space Program," given at the National Space Club in Washington, D.C., on February 17, 1972, Low partially defined his vision for the new NASA. He told his audience that although NASA's share of the U.S. budget had decreased from 4.3 percent of the total in 1965 to 1.3 percent in 1973, we could have a viable space program if we simply increased our productivity. He reminded them that the classical steps to improve productivity included investments in new tools and machinery, innovative techniques and modern technologies, skilled management and the training and diligence of the work force. NASA had increased productivity somewhat by using new techniques and technologies, but the real increase would come in the Shuttle era. The Shuttle would increase NASA productivity in a variety of ways; for instance, in terms of the 50 launch stands then in use, and 17 different combinations of boosters and upper stages, few would remain once the Shuttle was operating. The time needed to get space science payloads ready for flight would be greatly reduced, saving millions of dollars. And the experience of the Convair 990 flying laboratory aircraft program—which demonstrated a cost-effective, quick-reaction capability—showed the potential benefits the Shuttle would bring. Low closed his remarks with a reminder that the stakes were high and the challenges were great, but with the Space Shuttle, we could make it happen. To do it, Low declared, we must:

1. *Bring a productive Shuttle into being.* A productive Shuttle is one that performs as required, can be developed at a reasonable cost, and is economical to operate. If we meet the first two of these objectives, but not the third, we will have developed a white elephant.

2. *Provide for productive use of the Shuttle.* By this I mean that we must learn to take advantage of the special features that the Shuttle offers in our design of the things we will use in space. Experiments should look like those used in the Convair 990 aircraft, not like those used in the Skylab program. I am convinced that the person who is most innovative in making use of the Shuttle is also the one who has the best future in this business.

3. *Work together for the productive use of space.* NASA doesn't favor "manned flight" any more than it favors "unmanned flight." Our goal is to serve science and applications as well as the nation's need to explore and to maintain a presence in space in the best and most productive way. If we work together to meet this goal—as scientists and engineers, in industry and government—we will have a viable space program within our nation's means. If we don't, we will forego the challenge and the opportunity of America's future in space—and on Earth.

The Shuttle: A Reusable Spacecraft

Aviation Week & Space Technology magazine for Nov. 29, 1993, puts the cost per flight of the Space Shuttle between $463 million to more than $1 billion. The wide range depends upon how NASA chooses to count the costs.

14 Years to an RLV

"The July 2 announcement of the award of the X-33 contract to Lockheed Martin could be the most important space policy decision in almost a quarter century, since U.S. President Richard Nixon's Jan. 5, 1972, announcement that the United States would develop a partially reusable space shuttle.

"However, the award, using Neil Armstrong's words, is also just 'one small step' down the long path to finally creating what was hoped for in 1972—reliable, routine, low-cost access to space."

John Logsdon, "Small Steps Toward RLV Flight, Space News— July 8-14, 1996, p. 15.

Low's words on the Space Shuttle as a "white elephant," as regards economy of operations, proved to be prophetic.

Three months later, Low addressed an Aerospace Industries Association conference in Williamsburg, Virginia. In this address, "The Cost of Doing Business in Space—A Challenge to Business and Industry," he told his audience that increasing productivity in the space program had to be a joint venture of government and industry because 70 percent of the NASA budget was spent by industry. If we did not control cost, we would not only lose new starts but perhaps be forced to cancel programs already underway. He said that industry contractors had told him that one way to reduce costs was to leave them alone; that is, stop the detailed supervision of NASA projects by NASA personnel and reduce the paperwork. Low reiterated the need for a very close NASA-industry partnership.

He then made a statement that was to cause great conflict then and later, during the days of the operations of the Low Cost Systems Office. "And while I do want to improve the cost of NASA programs, I want us to achieve at the same time an even higher degree of success and not a lower one." Unfortunately, many of the NASA and industry employees whom Low needed to convince of his sincerity to reduce costs interpreted this statement as: Take no risk to reduce cost—it's business as usual—that is, 100 percent mission success at any cost.

Again, Low cited the Shuttle as a major tool for improved productivity and reduced cost, especially the reduction of payload costs. Since payloads had been designed for maximum performance and minimum weight, the cost of placing a pound into orbit was exceedingly expensive. But when launch costs were no longer the major expense, each Shuttle payload could be designed for low cost and high reliability. Low was convinced that if NASA and industry could capitalize on the huge cargo bay and heavy lift capability of the Space Shuttle, the cost of doing business in space would be drastically reduced. This, he believed, would be as great a technological achievement as anything else we had accomplished in space.

In 1972 there were 50 operating civilian spacecraft. In most cases each had developed a unique set of customized subsystems, such as telemetry, attitude control, command, propulsion and power. Each of these specifically tailored subsystems had required large amounts of money to develop. High costs were directly related to the number of spacecraft engineers needed to design those subsystems for just one requirement, the very high performance for the weight, the work within very tight tolerances, and the production in small batches on a short time scale. All this led to costly designs, enormous amounts of testing and the tightest of controls to ensure each subsystem was precisely within the designed and tested limits. Since the subsystems were only used once or twice, the designer went through the same process for the next spacecraft. Reliability suffered, due to low design margins and experience with any one item. Things were expensive because we built many one- or two-of-a-kind items and demanded the ultimate in performance from them. Low proposed to change this with the following formula:

First, we need to gain a better understanding of where exactly we spend the money. We need to identify those areas which have the greatest potential payoff for cost improvement.

Next, we need to determine standard ranges of requirements (for the systems or subsystems with the highest potential payoff) for our spacecraft of the future, so that we can develop a few "standard" systems instead of large numbers of individually tailored systems.

Third, we must see to it that these standard items are actually developed—each with a goal of low cost and high reliability. In other words, ultimately, these items should be available and listed in the catalog of preferred parts.

Finally, we must make sure that these items will be used—but I believe the competition of the marketplace will take care of that.[2]

In other words, Low focused first on hardware. Typically, hardware accounts for only one-fourth or one-fifth of the expense in a high technology spacecraft program. Nevertheless, Low reasoned, if the Space Shuttle design could accommodate a bigger payload, and if the payload could be much heavier than those in earlier spacecraft, then many hardware problems might solve themselves. Since NASA could relax the severe weight constraints under which it had operated, two things could happen. First, NASA could use a limited number of standard equipment items instead of those individually tailored for each flight, even though these standard items would have excess capacity for many missions. Second, the standard items would be designed with greater safety factors, have larger performance margins, be more rugged, and have the potential for improved reliability and cost. Improvements in cost and reliability would come from the economies of larger production runs, designs optimized for low cost, wider tolerances and therefore less individual testing and paperwork.

The Case of the Defective Tape Recorders

To further explain his vision, Low used his favorite example: tape recorders. NASA developed scores of them, at enormous costs, yet seldom completed a space mission without a tape recorder failure. He hoped to develop just two or three different types to meet all spacecraft needs. If weight were no longer a constraint, it might be possible to make them rugged enough so that they could be made more reliable, and at greatly reduced costs.

Tape recorder malfunctions had plagued NASA since its early days. In 1966, a Goddard Space Flight Center (GSFC) committee chaired by William K. Ritter was formed to investigate tape recorder problems on GSFC satellite programs. The committee's final report, dated May 12, 1967, was released by NASA for "Internal Use Only." The committee concluded that a clear picture of the problem was diffi-

---

Goldin Chastises U.S. Rocket Developers

NASA administrator Daniel Goldin harshly criticized the U.S. launch industry for using outmoded technology, designing commercial boosters that fail due to poor management and inadequate funding, and for having lack of vision in pursuing single-stage-to-orbit rocketry to reduce the cost of putting payloads into orbit.

"And then we launch junk," Goldin said. "We can't go on like this anymore. It's embarrassing. A great American company built a new launch vehicle and it failed. It didn't fail because of the laws of physics. It failed because of under-capitalization and poor planning," he said.

by Leonard David in Space News— February 12-18, 1996, p. 4.

cult to determine because of the lack of sufficient documentation, but most problems seemed to come from design, manufacturing or assembly rather than from changes in the technology. The committee's recommendations were to centralize responsibility for GSFC's tape recorders in one skill group, establish an interim group to work with GSFC project offices on uniform approaches to tape recorder development, and to make specific suggestions on design, procurement and testing. The committee report was not officially implemented, although a partial centralization of tape recorder technology was.

An onboard tape recorder.   *Photo courtesy of Lockheed Martin*

After yet another tape recorder failure on the Small Astronomy Satellite 1 (SAS-1), Low requested the Office of Advanced Research and Technology (OART) to review the performance of unmanned spacecraft tape recorders. A meeting of representatives from the Langley Research Center and the GSFC took place on June 17, 1971, and concluded that no improvement had occurred since the Ritter report, failures still appeared to be random, each project continued to develop its own tape recorders for its mission, and good engineering, rather than technical breakthroughs, were required to improve reliability and lifetimes. The group recommended that the Office of Space Science and Applications (OSSA) should act as the focal point for tape recorder improvements, and should allocate the necessary resources to develop several recorders for use by GSFC projects that could possibly become the nucleus of a NASA family of standard tape recorders. These recommendations were reported to Fletcher and Low, who asked OSSA to implement a "forceful plan" to begin the standardization of NASA spacecraft tape recorders.

Tape recorder improvement was also stimulated by the Planning Research Corporation (PRC) report on spacecraft reliability, 1958 to 1970. The report sampled 304 launches from 41 U.S. programs divided into two periods: 1958-67 and 1968-70. The major findings regarding tape recorders were:

The most failure-prone component appears, as it did in the earlier study, to be the magnetic tape unit with 38 failures occurring on 138 units observed. The failure rate for magnetic tape units in the combined sample (1958-70) is 40 failures per million hours—a significant increase over that reported in the earlier sample (1958-67) (28 failures per million hours). All other components have decreased failure rates compared to the rates reported earlier.

The publication of the PRC report and OSSA's slow pace increased Low's determination to move even faster on tape recorder standardization. He now asked OART to put together specific policy and plans for developing standard spacecraft tape recorders. The resulting Tape Recorder Action Plan Committee held its inaugural meeting at NASA in Washington, D.C., on January 21, 1972. After more meetings at the Jet Propulsion Laboratory (JPL) and GSFC, the committee reported the following conclusion and recommendations to Low on March 21, 1972:

Findings from the Tape Recorder Action Plan Committee

- Tape recorders are the most failure-prone component in U.S. spacecraft.

- The recorder is a sensitive and delicate mechanical mechanism affected by operating and storage time and the environment to which it is subjected. It has been constrained to work on minimum power, at a minimum weight, and at minimum size.

- The small quantity of tape recorders produced for each separate project and the resulting restricted amount of test time contribute to reduced tape recorder reliability. . . .

- Emphasis on spacecraft system contracting and the project form of management structure has caused a heavy dependence on industry for tape recorder technological support. This has resulted in a large number of different types of tape recorders being developed for the various NASA projects.

- The amount of recorder technology transfer from project to project has been generally inadequate because of the lack of strong continuous participation by tape recorder discipline or skill groups within most NASA Centers along with the absence of inter-Center communications and coordination in this field.

- A single lead Center at the component level of tape recorders is not appropriate because each project office and Field Center must retain the ultimate responsibility for the success or failure of its assigned missions and the components required to support them.

Recommendations

- Increased level of support and technical direction to each project should be provided within each Center from the tape recorder skill group so that experience from one project can be effectively transferred to another.

- When spacecraft system contracts are used, strong consideration should be given to providing the tape recorder as government-furnished equipment.

- A limited set of standardized tape recorders should be defined that will satisfy most known manned spacecraft requirements.

The tape recorder history, along with the planning for the Shuttle utilization, inspired Low to move out now on his Agencywide Low Cost Systems initiative, and he had an ideal first candidate for equipment standardization in the trouble-plagued tape recorders.

On May 29, 1973, Low assigned responsibility to the Goddard Space Flight Center for the development and procurement of tape recorders in the 10^8 and 10^9 bit storage range. These tape recorders were the first of the NASA standard equipment developments.

Forming a Cost Improvement Task Force

Low let everyone know just how serious he was about improving costs by establishing a task force reporting directly to himself. Low sent a memorandum to all NASA officials on May 16, 1972:

SUBJECT: Space Vehicle Cost Improvement

The high cost of doing business in space, coupled with limited and essentially fixed resources available for space exploration, places severe limitations on the amount of productive work that NASA can do, unless we can develop means to lower the unit cost of space operations. It therefore becomes an item of first order business for each of us to find ways to drastically reduce the costs of all elements of space missions.

I am convinced that major cost improvements can be realized, and that this matter should become a first order of business for all of us. A basic approach to lowering the costs of space systems should include the following:

Typical Tape Recorder Project Costs
($ thousands)

Program	Development Costs	Cost per Recorder
Explorer (AE/C, D)	750	83
Nimbus E	500	125
Orbital Solar Observatory F&G	800	80
Apollo	8300	80
Mariner H, I	3166	250

Anticipated Standard Tape Recorder Savings
Millions $

Sources:
1. Report of Tape Recorder Action Committee
2. Standard Tape Recorder Execution Plan
3. Goddard Plan For Magnetic Tape Recorders

Savings = $11 Million

Current Recorder Cost

Standard Recorder Cost

Years

U.S. Senate Appropriations Testimony, March 13, 1974.

1. A detailed understanding of exactly where we spend our money.

2. The determination of range of requirements (for the systems or subsystems with the highest potential payoff) for our spacecraft of the future.

3. The development of "standard" systems or subsystems, designed for low cost and high reliability. (We need a catalog, ultimately, of available preferred parts.)

4. A method for assuring that as a rule only the "standard" systems are used.

I consider this effort of such high importance and priority that I am prepared to devote whatever resources are required, both inhouse and on contract, to achieve significant results.

George Low and Dr. James C. Fletcher, NASA Administrator, during a Space Shuttle briefing at San Clemente, California, January 5, 1972.

Photo courtesy of NASA

I am hereby establishing a task force, chaired by Del Tischler, to carry out steps 1 and 2, above, and to develop a plan, goals and objectives for steps 3 and 4. I want each of the offices to provide the necessary support to the task force, especially in terms of experienced people.

The task force is authorized to place requirements on the various line organizations to accomplish its objectives.

Footnotes

[1] "The Next Decade in Space," A Report of the Space Science and Technology Panel of the President's Science Advisory Committee, March 1970.

[2] George M. Low, "The Cost of Doing Business in Space—A Challenge to Business and Industry." AIA Conference. Williamsburg, Va., May 20, 1971.

Chapter 3

Let's Get Fresh Ideas on Cost Control

To expand his basic vision of cost control at NASA George Low began visiting various industries. On June 13, 1972, he went to the Fairchild Hiller Corporation in Farmingdale, N.Y., and then on to the Westinghouse Corporation in Pittsburgh. At Fairchild Hiller, Low was briefed on the AX Close Support aircraft then under design for a flyoff competition with a Northrop Corporation aircraft. A unique feature of this competition was the Air Force establishment of goals rather than requirements; the goals could be traded off against costs. New development efforts on the aircraft were minimized to cut costs and development risks. Managers at Fairchild Hiller were careful not to overdesign, and they used innovative means to simplify manufacture. They stressed simplicity, used a minimum number of parts, and involved the shop planners in the design phase. When the subassembly designs were completed, they conducted a detailed cost analysis on each in order to compile a realistic bid on the 600-unit aircraft production order.

AX Close Support Aircraft, later the A-10 of Desert Storm fame.   *Photo courtesy of DoD*

In response to a direct query by Low on how costs on NASA programs could be reduced, the Fairchild staff said that NASA frequently got too involved at the subcontractor level, causing the prime contractor to lose control and allowing costs to grow. They thought NASA should use more fixed-priced contracts or establish firm cost ceilings and let the performance come out where it may. The Fairchild Hiller employees felt the Advanced Technology Satellite (ATS) they built could have cost up to 25 percent less if NASA and the contractor had initially agreed to a realistic cost target and the importance of holding to that cost, and if NASA had not overspecified.

At Westinghouse the group toured the Cost Improvement Laboratory. The laboratory consisted of a single room with table and blackboards and a staff of five engineers, each with a different specialty—design, manufacturing, control, etc.—who undertook a cost improvement analysis on projects coming from the operating division. The projects to be analyzed had to come with a detailed cost breakdown. The laboratory staff acted as the catalyst during the analysis, with the project staff doing the work. This process worked because cost visibility was in an understandable form, enabling them to attack the high cost items. Their average cost improvement was 30 percent. The basic formula was to make cost visible, put the team together to generate alternative ways to perform the function, identify savings, and combine the best alternatives for the greatest payoff. As they reduced the number of parts, they usually also improved reliability.

On July 11, 1972, Low visited Warwick Electronics, a subsidiary of Whirlpool and the primary supplier of television sets for Sears Roebuck and Co. Warwick's success was based on standardization and commonality. They used only two chassis for all Sears TV sets and had started a standard piece parts program. The company reduced overhead by laying off mediocre people. In establishing the cost of a set, they worked to a tight target cost, rather than designing to "lowest possible cost." Warranty costs were paid by Warwick, not Sears; therefore, there was ample incentive to keep quality a prime objective. With respect to competition, they bought all competitors' sets and tore them apart to see what they could learn.

That same day he visited Ford Motor Company, where he was briefed on their absolute cost control process, the main feature being a detailed understanding of all cost elements involved in the product. Each division was assigned firm cost objectives and was responsible to find offsets if an "uncontrollable" increase should occur. Top management's interest and prodding helped keep costs within the estimate. They kept their engineers involved in meeting the cost targets. Ford personnel emphasized the importance of monthly meetings with the Executive Vice President, who reviewed their cost performance. Ford also bought their competitors' cars, took them apart and conducted cost estimates on them. They bragged that they knew exactly what a Chevy should cost.

The main points Low took away from Ford were:

1. All costs must be known and clearly understood.

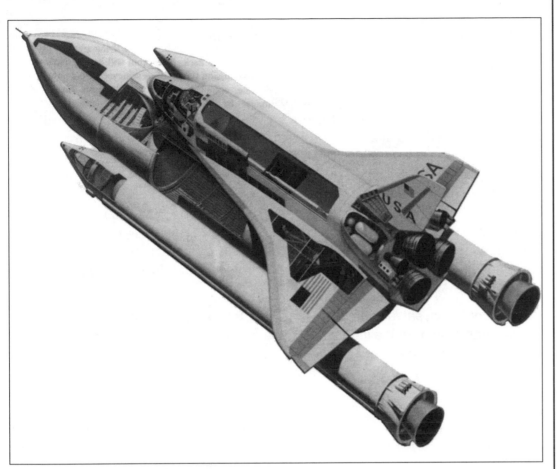

Space Shuttle cutaway, circa 1975. *Photo courtesy of NASA*

2. Firm cost targets must be established (as opposed to seeking "lowest possible costs").

3. The cost control system must be managed by responsible people throughout the organization.

4. There must be high interest and visibility by top management.[1]

Applications to the Space Shuttle

On July 31, 1972, Low addressed the Shuttle Sortie Workshop and told his audience that the gloom which seemed to surround the space program was unwarranted. Although the budget was less than desired, there were things that could be done about reducing the costs of doing business in space, and that would allow NASA to do more within the budget constraints. Low cited three ways to do this:

1. Take advantage of the relatively unconstrained weights and volume available on the Shuttle,

2. Optimize for low cost and high reliability, and

3. Use standard equipment to the maximum extent possible.

Low reminded the audience of the cost task force he had recently established and his industry visits; of the AX Close Support aircraft competition, a no-frills aircraft to be produced at a low target cost; of the establishment of performance goals but no performance requirements; of Westinghouse's Cost Improvement Laboratory and its outstanding record in reducing costs; and the absolute cost control system of the Ford Motor Company.[2]

He told his audience that he had yet to pull this all together as a new way of doing business in NASA, but that he had a few observations. These observations may appear similar to previous ones, but the subtle differences reflect his better understanding of the dimensions of the cost problems and the potential solutions available to NASA.

1. *Don't reinvent the wheel.* Use the best that is available from other programs. In all of the industries I have visited, "not invented here" is unheard of. All tear down their competitor's product, study it, analyze it, cost it, and make use of the best ideas in it so long as they do not violate patent rights.

2. *Standardize.* This applies to parts, components, modules, subsystems, and entire systems. Warwick Electronics has only two different chassis for its entire line of TV sets; and the left and right landing gear on the A-10 are identical!

3. *Design to minimize testing and paperwork.* Note that I did not say "minimize testing and paperwork"; I said design to achieve this. Simply stated this means: use larger margins and higher safety factors. In Apollo we spent million of dollars—on tests and paper—to be sure we did not exceed the "fracture mechanics" limits on our pressure vessels. A few extra pounds in tank weights would have completely eliminated that problem, and the testing and paperwork along with it.

4. *Know your costs.* None of the things I have said so far has any meaning if you don't know how much each element costs. The area of accurate cost estimating is one where we have a great deal to learn.

5. *Trade features for cost.* This follows naturally from the previous item. Once we know how much something costs, then we can ask ourselves whether it is really worth it. Many of our so-called "requirements" really aren't that firm, and should be stated as "goals," to be re-examined in terms of cost.

28

6. *Pay particular attention to the few very high cost items.* In many designs some small percentage of the items amount to most of the costs. By knowing the costs, and by listing items in order of descending costs, it becomes possible to devote a great deal of attention to the high cost items—generally with profound results.

In the implementation phase, I would emphasize the following points:

1. *Know your costs before you start.* This perhaps is the most fundamental of all requirements. Without exception, the NASA programs which have been in difficulty were the ones that had insufficient definition at the outset.

2. *Set firm targets.* A desire for the "lowest possible cost" is not a good way to approach the job. A firm and absolute cost ceiling should be established for each job.

3. *Meet the established cost targets.* Don't blame cost growths above target on "external forces." Find ways to meet the targets, no matter what happens. This means that you have to become more productive in one area, if another area exhibits an "unavoidable" cost increase.

Low then introduced a new ingredient in the cost equation, one that critics of NASA previously said was totally missing—the customer.

Let me change the subject now, and briefly talk about another important area: user involvement. This workshop is a good example of what I have in mind. You are here to discuss jointly what the Shuttle—in the sortie mode—should be.

I particularly want to remind those from the manned space flight organization who are participating in this workshop that the only reason for developing a Shuttle is to provide a service to all potential users. If it won't do that, then there is no point in building it.

This may require a new attitude on the part of some of us at NASA. Specifically, we must learn to give the user what he wants, and not what we think he should want! I am sure that this workshop is the right first step in this direction.

Low's next address, "NASA's Attack on the Cost Problem," was given at the National Security Industrial Association and the Armed Forces Management Association Symposium on Cost: A Principal System Design Parameter. In this address he recognized the differences and similarities between NASA and DoD. DoD differed from NASA in acquiring systems in quantity while NASA generally bought in lots of 10 or less. Therefore, DoD placed a high level of importance on production and operational costs; NASA on development costs. The similarities were that NASA and DoD were involved in costly complex systems frequently pushing technology development and were both facing major cost problems. Low was blunt in describing the urgency and need NASA faced in reducing the high cost of doing business in space. He described how the Agency was on the verge of exciting new discoveries and applications but could not move out rapidly because NASA simply could not afford it. Low also cautioned that this nation could lose its hard-won world leadership in space if we did not find a way to get more for our money.

Low then told the DoD group what he had found out about how to reduce costs. He had visited the company that made television sets for Sears and Roebuck and learned how they increased productivity and beat the foreign competition. He had been to Detroit and talked to the auto industry about plans to develop and produce a successful small automobile. He saw how a major appliance firm reduced cost by 25 to 30 percent. He described his experiences flying on a remote sensing mission in a Convair 990 flying laboratory aircraft used by NASA for science and application research. He described the mission in some detail, noting that 12 investigators were on board using a variety of highly sophisticated complex instruments conducting 14 individual experiments. Low was impressed with the overall conduct of the mission, but especially with the ease and simplicity of it. This mission was conceived and flown in five months. Each investigator was given a standard rack which would later be fitted into the aircraft. With the rack came instructions regarding power, data and weight limits. If the experiment could be accommodated within the limits of the standard rack, no other test or paperwork was needed. He remarked that, in his estimation, a similar mission flown into space would take a very large team of people scattered throughout the country with stacks of paperwork, countless revisions and meetings, rigorous tests and enormous sums of money.

The investigators in their laboratories assembled their instruments and arrived with them at the Ames Research Center two weeks before the scheduled flight. The equipment rack was installed in the aircraft, inspected for workmanship, attachment and interfaces and, after a local checkout flight, the aircraft was ready for the mission. The next day the aircraft departed for the west coast of Africa to rendezvous with a number of Soviet surface vessels in a pre-planned international program.

As impressive as the quick response was, even more impressive was the cost comparison. The cost of developing a spectrometer for the aircraft mission compared to the cost of a similar instrument developed by the same investigator for use on Apollo 17 was $100,000 to $3,500,000. The paperwork alone at the investigators' laboratory for the Apollo instrument was $500,000. Low, as the former Apollo Spacecraft Program Manager, was reluctant to say things should have been done differently on Apollo, but he wanted to show there were very large savings to be made on programs that did not have the tight constraints of Apollo.

A rebuttal to George Low's comparison of the spectrometer developed for the CV-990 experiment and a "similar one" developed for the Apollo Program was

The CV-990 over Monterey Bay during checkout flight for instrumented nose boom.

NASA CV-990 Airborne Laboratory
Since April 1965, NASA had been operating a Convair 990 aircraft for research activities in atmospheric and space sciences, applications technology and aeronautics. The flying laboratory was based at Ames Research Center (ARC), Moffett Field, CA. It was operated for the benefit of researchers whose proposals had been approved by NASA program offices. Airborne laboratory flights could be conducted out of Moffett Field or from remote bases worldwide, according to scientific requirements.

The CV-990 was a four-engine jet transport aircraft with a range of about 3,000 nautical miles, a ceiling of 41,000 feet (12,500 meters), and an experiment payload of 20,000 pounds (9,070 kilograms). Its utilization had been 400-500 flight hours per year. Special viewpoints, power systems, and instruments had been installed to support a wide range of research programs.

The Spacelab-type racks built for the CV-990.   *Photos courtesy of NASA*

31

quick in coming from the Associate Administrator for Manned Space Flight, Dale Myers. His response pointed out that CV-990 spectrometer was 100 times less sensitive than the Apollo instrument and did not operate on the wave length of interest to the lunar science community. The high reliability required for a single lunar mission could not be met by the CV-990 instrument. This instrument failed on its first flight even with its principal investigator as the instrument operator. Other factors included the space hard wiring requirement, and the weight and volume constraints of the Apollo mission. Myers' response ended with a comparison of a spectrometer that had flown in space on Mariner 69. This instrument cost NASA $2,848,000.

Low next shared his updated conclusions on what needed to be done to reduce cost. He cautioned that this was a progress report, in that he and NASA were still learning. He split his observation into two categories, design and implementation. Note the gradual evolution of his previous listing from a few months earlier.

In the design phase:

1. *Don't reinvent the wheel.*

2. *Standardize.*

3. *Design for low cost.* Involve production engineers in the earliest stages of design to help eliminate those things that will be difficult to produce.

4. *Design to minimize testing and paperwork.*

5. *Recognize that different systems can accept differing degrees of risk.* In the example I cited a few minutes ago, the Apollo spectrometer must have a much higher assurance of working than the same instrument flown in the CV-990. A Shuttle life support system must be more reliable than a simple experiment in the Shuttle payload bay. The cost of a system should reflect the acceptance of risk in those instances where this is possible.

6. *Know your costs.*

7. *Trade features for cost.*

8. *Pay particular attention to the few very high cost items.*

In the implementation phase, I would emphasize the following points:

1. *Know your costs before you start.*

2. *Set firm targets.*

3. *Meet the established cost targets.*

In summary, we must find ways to design for lower costs, we must know our costs, and we must set out to meet those costs. This works in successful firms in the commercial world, and there is no reason why it shouldn't work on Defense/NASA programs as well.[3]

Unfortunately, the CV-990 aircraft mission gave Low a false sense of what would be possible on the Space Shuttle.

Continuing his industry meetings, on November 3, 1972, Low attended a Design-to-Cost Symposium at Boeing in Seattle. He was impressed by Boeing's ability to become "lean and mean." Boeing had reduced its Headquarters work force from 2,000 to 200 people, thereby forcing a management-by-exception mode. Boeing executives repeated time and again the increase in productivity achieved by moving Headquarters people to work in divisions. Their thought was if Headquarters people were good, they should be in the field, doing; if not, they shouldn't be at Boeing at all. They had eliminated the executive dining room and executive aircraft, they reduced their corporate planning staff from 130 to 20 people, placed all of their laboratories under a single manager and all of their facilities under another single manager. Their aerospace group had done away with organization charts. Low was disappointed that the presentations did not show how Boeing designed to cost but rather concentrated on how Boeing learned to meet cost objectives.

About this time Low shared with Del Tischler his thoughts on a low cost guidelines paper Tischler was preparing for widespread distribution in NASA.

I would suggest, however, that you try to take a second cut at a guidelines paper, which at the very top level, would be split into two separate sections—one on management approaches and the second on implementation.

On management approaches, the best guideline I could give is to run a "tight ship," one that is "lean and mean." This means that at each level in the organization—Headquarters, Centers, contractors, subcontractors— there are only sufficient people so that they are all "doers" and there are no "hangers on.". . . These project people . . . should be supported by highly competent additional technical teams which can be called in whenever the need arises.

In a "lean and mean" organization there is an additional responsibility for each level of management to immediately report problems to the next higher level. If this responsibility is known to all managers and they know that their job depends on this, then one can immediately do away with all sorts of levels of checks and balances, over-the-shoulder lookers, etc.

In the implementation phase I am beginning to see that there are two important factors which we must consider. They are "design to a lower cost" and "manage to meet the cost objectives." . . .

I feel that the most important job that NASA must now learn how to do is to design to a lower cost. . . . The second part, managing to meet cost objectives, is one where I do believe both industry and government are becoming much better. . .[4]

As 1972 drew to a close, Low was beginning to understand some of the implications to the industry and to NASA of converting to a low cost culture. He was disappointed in not having a more cost-effective, fully reusable Shuttle to work with, but was still convinced that major cost reductions could and must be made.

Low continued his information gathering visits during the new year, 1973. But now he wanted to concentrate on the NASA project management work force. He felt strongly that this group, the "doers" as he described them, would be the ones to make happen "what I have been preaching for the past eight or nine months." In late January he flew to the West Coast with Del Tischler to visit with members of the Ames Research Center (ARC), Mountain View, California; the Jet Propulsion Laboratory (JPL), Pasadena, California; and the Dryden Flight Research Center (DFRC), Lancaster, California.

At ARC, based on their experience in managing airborne science and the Pioneer and Biosatellite programs, the group concluded that big projects are hard to control. Also, they believed that project scientists and engineers must be close-coupled; it's helpful if the hardware producing company is close by for quick, just-in-time problem solving. NASA should not make a big ritual of experiment selection; avoid at all costs an initial work force build up; and "generators" should be separated from the "doers." The contractor's paper system should be used rather than inventing one inhouse; project objectives should be limited to one or two big ones instead of many small ones; and, lastly, one of the most important, simple interfaces are essential. In Low's words, "the basic message is: keep things small and keep them under control."

At JPL, Low ran into an adversarial situation with a group of managers who seemingly had not thought about the cost problems as much as the ARC managers had. But two people, Bob Parks and Gene Giberson, provided valuable cost control insights. Parks, the Assistant Lab Director for Projects, gave a report on potential standardization by block spacecraft, and Giberson, the Project Manager of Mariner Venus-Mercury, reported on his project. Low was very impressed with Giberson's management style. Giberson had his project well under control and believed the best cost control tool was the fixed-cost ceiling imposed on his project. The ceiling helped him establish tight cost controls and also provided his managers with the basis for a very different attitude. It was difficult at first to get his managers to believe they could manage to the cost ceiling, but once they did, they did an admirable job. He could not recall any decision he made due to the cost ceiling that was detrimental to the project.

Giberson strongly recommended, however, that where a total project cost ceiling is imposed, NASA should not then impose an annual cost ceiling on the project. This annual ceiling would take away the flexibility needed to manage to the total ceiling.

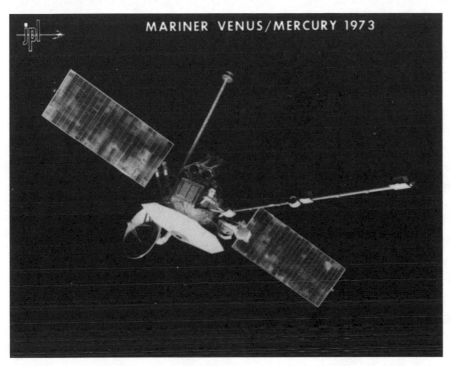

Mariner Venus/Mercury 73 Spacecraft. *Photo courtesy of JPL*

The DFRC staff had given a great deal of thought to the problems of reducing costs and were convinced that the best approach to low cost was the experimental shop approach. George Low was not sure of this, but agreed to sponsor a future NASA/industry experimental shop symposium. Low thought that the entire Center was an experimental shop and that they had a great deal of fun running their programs.

Keeping up a rather brisk pace, Low visited the Lewis (LeRC) and Langley (LaRC) Research Centers on January 30, and February 1, 1973. At LeRC, the flight projects (with the exception of Centaur) were small and consumed only a small portion of the Center's resources. However, project managers had use of the entire Center work force when necessary. The LeRC project managers believed tight cost ceilings to be absolutely essential in running their projects.

The experimental shop approach was in use for the D-1 Centaur, and the LeRC people were convinced this was the way to go on all the projects. They felt this approach could be used on any size project because a large project could be broken into a number of small experimental shop packages. The Center Director, Bruce Lundin, told the story of the RL-10 engine. He did a detailed analysis on the cost to purchase a single piece of tubing for that project. The price of the labor and materials was $29, but by the time all the other "non-productive" effort was added,

Cutting Space Station Costs is Now up to Boeing

The new funding level says people get off the payroll . . . it's not the hardware, it's the people.
– Bryan O'Connor

Space News—
August 23-29, 1993,
p. 7.

the cost had risen to $250. This convinced him to renegotiate the contract with Pratt and Whitney from a Cost Plus Fixed Fee (CPFF) to a Fixed Price (FP). He closed the resident office and in general got out of the contractor's hair. The original cost estimate under the CPFF contract was $600,000 for each engine—the actual price under the FP contract was $340,000. He estimated the contractors still made 20 percent profit on each engine.

At Langley, Low reviewed the management experiences on the Dual Air Density Satellite, the Scout Launch Vehicle and the Viking Mars Lander. The bulk of the discussion was on Viking. Jim Martin, the Viking project manager, strongly believed the only way to manage a project as large as Viking was in the Apollo mode. He did not believe the experimental shop approach could work on a large project. He also felt that some money (in Viking, 1.5%) should be allocated to the selected contractor before freezing requirements and fixing the target cost. Martin believed that having an agreement on a fixed price with Headquarters was important in managing Viking.

At the Kennedy Space Center, the discussion centered on the responsibilities of the line and functional organization. The group also talked about the way NASA and COMSAT did business. There was an agreement that COMSAT brought a performance specification to a contractor, then left the contractor alone, while NASA was still working at the experiment level and was much more involved than it should have been. NASA might also be overdoing the number of people, paperwork, and checks and balances. The Center Director, Kurt Debus, suggested that cutting back on people until it hurts may be the only real key to reducing costs.

On February 12, 1973, keeping up an aggressive pace, Low and Tischler visited the Goddard Space Flight Center (GSFC). Low first met with the managers of the Explorer class satellites. Many of these satellites were low cost products with common subsystems developed and assembled inhouse. However, there was no commonality with other GSFC spacecraft or spacecraft of other Centers. GSFC management felt strongly that tremendous savings could be realized if some subsystems were bought in bulk and stored until needed on future projects. This would require additional upfront funding.

Low next talked with the sounding rocket people. The real key to keeping costs low in this program was the opportunity to refly if they failed, usually within nine months. With this assurance, their management processes were greatly simplified. Low also reviewed a typical inhouse satellite payload project which he described as GSFC's form of an experimental shop approach.

Low then met with Joseph Purcell, who had pioneered the design of standard subsystem modules, typically used for propulsion, guidance and power. Purcell believed that NASA's future challenges should come from flying the best possible instruments, not from building better spacecraft. He thought using standard modules on projects already approved would save the Agency and its customers more then $500 million.

In the afternoon, Low met with the remaining project managers and other senior Center managers. As Low would later relate, the Center managers were still

36

very bitter over the failure review he had ordered on the Delta Launch Vehicle failures and questioned if such a review would be called every time they had a failure. If this were the case, they would take no risks to reduce costs. Low assured the audience that if a project manager took a higher risk to save money and the risk was communicated to management, he would back them all the way, especially in case of failure. However, he would not back an incompetent project manager.

Another important topic was the tradeoff between the experimenter's requirements and the real cost resulting from them. Discussion clearly showed that there was very little requirement tradeoff going on between project engineers and scientists. When Low questioned the group as to what NASA requirements drove costs, the almost unanimous answer was systems safety. Someone then asked Low how many low cost failures he would tolerate. It was obvious the question bothered him, but after a short pause he blurted out, "One out of five." (After the visit Low rode back to Headquarters in silence. Suddenly he turned to his assistant and said: " I shouldn't have said that." "Said what?" asked his companion. "I shouldn't have said 'one in five.' I don't want *any* failures.")

A Goddard project manager told the story of his discussion with a contractor about a future satellite project. The contractor asked how NASA would want them to bid the job: the traditional NASA way or the Air Force way or the COMSAT way. In the latter two, the bid would be about 20 percent less than for NASA. Another manager told of a discussion with the Air Force people who managed small projects. The Air Force had a level-of-effort research and technology satellite program (Space Test Program) where they attempted to get maximum experimentation for a fixed amount of money.

From all these discussions it was clear to Low that the amount of risk taken in a development project was directly related to the ability to refly the experiment at an early date. Thus, to reduce costs it might be necessary to take higher risks and, therefore, NASA should budget enough money to refly experiments that failed.

Low prepared a separate memorandum on a proposal he received from a GSFC project manager. The proposal pointed out that single spacecraft missions cost much more than they should because project managers are buying "insurance" for first flight success due to the fact that it was virtually impossible to schedule a refly. Low wanted to seek practical solutions for this dilemma.

Low and Tischler visited the Marshall Space Flight Center (MSFC) and the Johnson Space Center (JSC) on February 22 and 23, 1973, respectively. At MSFC they reviewed the High Energy Astronomy Observatory (HEAO) project, the Shuttle and the Shuttle Main Engine project. HEAO was designed to use 80 percent off-the-shelf equipment, but much of the equipment had to be modified. Regarding the Shuttle, MSFC pleaded for simple and standard experimenter interfaces similar to the CV-990 program model. MSFC had examined the current mission model and estimated that a standard spacecraft could accommodate 43 missions, thus reducing 32 different subsystems to just nine.

At JSC, Low met with more than 30 members of the project management community. Initial discussion centered on the Shuttle, and all agreed it must be safe and reliable. Cost savings would come from payloads using off-the-shelf equipment, and while payloads would be safeguarded, reliability would be left up to the experimenters—a new concept for NASA. An observation was made that simplicity was the key, and when we learn to do things simply, they will also be cheaper.

On July 30, 1973, Low visited the Fort Worth plant of General Dynamics to look at the YF-16 lightweight fighter prototype. The YF-16 was one of two lightweight fighters, the other being developed by Northrop. Both companies had contracts to produce two prototype aircraft. The total value of the two contracts was $74 million, with approximately half going to each of the contractors. The two fighters would be involved in a flyoff competition. The winner would be assured of decades of Air Force orders.

The General Dynamics F-16. *Photo courtesy of DoD*

The project management team was quick to point out that it required a "cultural change" on the part of the Air Force to be able to do many of the things they were then doing. The Request for Proposal for the fighter was just two pages long, and the proposal was 60 pages long, 50 of which were technical and 10 devoted to management. In addition to the proposal, the contractors had to submit wind tunnel models which were tested in NASA facilities. Convair had spent $5.6 million of corporate money before proposing.

There was no statement of work for this competition, nor were there any externally imposed specifications. The competition with Northrop took the place of both of these. However, there were top-level goals, covering turning rates, accelerations, range and speed.

Without required formal Air Force documentation, Convair ran the project using their own documentation. The project team made their own decisions without engaging in lengthy debates and discussions with the Air Force. Cost was first among equals in each design decision.

Footnotes

[1] George M. Low, "Visit to Ford Motor Company," memorandum for the record. Washington, D.C., NASA, July 19, 1972.

[2] George M. Low, keynote address, Shuttle Sortie Workshop, July 31, 1972.

[3] George M. Low, "Cost – A Principal System Design Parameter." Washington, D.C., August 16, 1972.

[4] George M. Low, memorandum to A. O. Tischler, November 13, 1972.

Chapter 4

We Had It About Right the First Time

When he appointed Del Tischler to head up the Space Vehicle Cost Improvement Task Force on May 16, 1972, George Low got a respected technologist with experience in project management and large systems development.

Tischler had joined the Lewis Laboratory of the National Advisory Committee for Aeronautics (NACA) as a chemical engineer in 1942. Tischler had been happy at LeRC doing research on fuels for various kinds of engines, racing sailboats on the Great Lakes in the summer, skiing in the winter, pitching his NACA team to two championships in the Ohio Industrial softball league. Meanwhile he had switched from research on fuels for the then-new jet engine aircraft engines to rocketry. While the aeronautics-oriented NACA virtually disdained rockets, Tischler's work in the obscure field of combustion-driven oscillations won him wide acclaim.[1]

One Saturday afternoon in June 1958, Abe Silverstein, then Director of the Lewis Laboratory, called to say: "I want you at NACA Headquarters in Washington on Monday." There, among such soon-to-be legendary figures as Bob Gilruth, Max Faget and Ed Cortwright, Tischler helped lay the plans to make space history. Because the NACA Headquarters building next to the Dolley Madison house on Lafayette Square was crowded with new people who were forming a space agency, Tischler occupied former NACA Director Jimmy Doolittle's office on the top floor next to the auditorium and sat in Doolittle's enormous leather chair behind a big oak desk. Tischler was to move that chair many times in his career.

Del never swung a softball bat again. Within days after the formation of NASA on October 1, 1958, he initiated the F-1 engine development (NASA's first major contract) and the first component of the Apollo program. In 1961 he was assigned to be Assistant Director, Launch Vehicles and Propulsion, Office of Manned Space Flight (OMSF). In 1964 he was appointed Director, Chemical Propulsion Division Office of Advanced Research and Technology (OART), where he was responsible for technology development for both liquid and solid propellant rocket motors.

In 1969 he was appointed Director of Shuttle Technologies Program, a multi-disciplinary Agencywide program that he organized and ran as a joint OMSF/OART effort while retaining his Propulsion Directorate and attending courses at the Harvard Business School. This work won him two Exceptional Service Medals from NASA.

Thus Tischler knew technology, program management, NASA, and the aerospace industry. He had been elected a Fellow of the American Institute of Aeronautics and Astronautics, had won many awards and was a member of several interagency planning and coordination committees. When George Low appointed Tischler he got the ideal candidate for a most challenging job.

In June 1972 Tischler wrote Low of his intentions on organizing and implementing the effort. Tischler believed the main issue was a major change in NASA's deeply ingrained thinking on space program management from cost as a minor goal to cost as a principal goal and possibly, the limiting parameter. Tischler was convinced that redundancy was a major part of the problem. Redundancy in hardware—coupled with redundancy in engineering, management, inspection, testing and reporting—all had to be re-examined. Opposition, Tischler believed, would come when new procedures or processes threatened old organizations and jobs. Tischler's plan of attack was to form a steering group under his leadership. The steering group would consist of panels and subpanels devoted to specific functions. He would solicit help from the NASA Field Centers in organizing and operating the panels. The Steering Group would identify cost drivers in the present programs, identify common space systems with large future cost reduction potential, develop "standard" hardware specifications, prepare a catalog of preferred parts, and help formulate rules to ensure compliance to a yet-to-be developed low cost process.

Tischler felt that he had to concentrate on:

• Standard hardware definition and development

• Management guidelines covering transportation, carriers and experiments

• Simplified management procedures, and

• Demonstration that low cost payloads were practical.

To be really successful, NASA must include "customers" in this effort, the science and research communities. In addition, Tischler had a host of procedural questions yet to be worked out. He closed this communication by sharing with Low an emerging concept of a "standard equipment store."

A computer-maintained equipment inventory was to be managed by NASA, and although no actual hardware was to be "stored" in the store, its inventory would be equipment descriptions and specifications, location of existing units, manufacturer, per unit cost, mission uses and performance reports.

The "store" responsibilities included: mission surveys, systems analyses, requirements authentication, coordination of specifications (mission acceptance), module (subsystem) development, a list of test equipment, checkout procedures, lists of software, simulations, a catalog, a computerized inventory, research and product improvement, and enforcement of standard system use.

The store would be managed by NASA personnel, independent of the Mission Office Program. This concept was to provide a framework for the later Standard Equipment Program.

At a meeting of Tischler's task force in late 1972, the full extent of the complexities of the cost problem were addressed for perhaps the first time. The group reported that in order to reduce payload costs by a factor of 10 (a self-imposed early objective), NASA would have to change procurement practices, management philosophies, design approaches and operation methodologies.

In studying 13 different payloads, the group found that design, development, test and engineering amounted to 72 percent of total payload cost, and production the remaining 28 percent. After quizzing laboratory and project managers as to the past management emphasis, the group found technical performance was the main goal of over 60 percent of the respondents, with cost performance less than 5 percent.

In attempting to explain cost increases, the group found a broad range of factors, from more than 45 percent of the increases directly related to definition of requirements, to less than 5 percent blamed on inaccurate cost estimates. In examining management practices that drove cost to the higher end of the scale, the group identified five as major cost drivers:

1. Traceability requirements

2. Redundancy

3. Test and verification requirements

4. Extensive pre-launch checkout process and procedures

5. Development of backup payloads

The group then considered what was needed to improve past cost performances and agreed that there must be four major thrusts:

1. Hardware standardization

2. Emphasis on design for low cost

3. Reduction of documentation/procedures

4. Use of demonstration projects

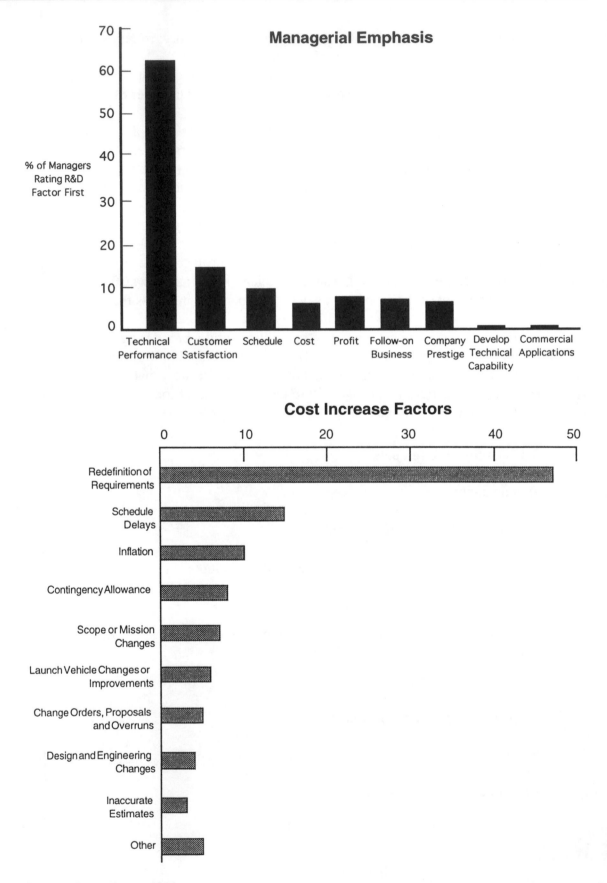

Managerial Emphasis

% of Managers Rating R&D Factor First

Technical Performance · Customer Satisfaction · Schedule · Cost · Profit · Follow-on Business · Company Prestige · Develop Technical Capability · Commercial Applications

Cost Increase Factors

Redefinition of Requirements
Schedule Delays
Inflation
Contingency Allowance
Scope or Mission Changes
Launch Vehicle Changes or Improvements
Change Orders, Proposals and Overruns
Design and Engineering Changes
Inaccurate Estimates
Other

Steering Group Report 1972.

The group's insight in payload development was prophetic: "Aerospace managers (NASA as well as contractor) must become attuned to 'assembling' space systems in contrast to developing them."

The group next looked at the "needs" of the NASA project managers to manage costs better. They concluded that a well-defined project plan, credible cost targets, cost incentives, analytical tools for low cost management, low cost technology support, and a redirection of NASA's "success" syndrome toward low cost were all important needs. In addition, they recognized a universal need for education in low cost technology, design and management, decision theory and risk assessment.

To optimize cost, the group thought NASA must conduct tradeoff analyses between development cost, development goals and development risk. But this process would require a new management methodology for making tradeoff decisions. This new methodology must integrate decision control features of systems engineering, value engineering, risk assessment and decision theory, and the tools would have to be developed by NASA. The panel also warned that the transformation of an engineer into a cost-conscious professional manager was not an automatic consequence of the assignment, but one that required training.

As to the black art of cost estimating, the panel recommended a more consistent Agency project cost accounting process and the adoption of a standard NASA work breakdown structure. To improve NASA's inhouse capability, the Agency must conduct special training, including the latest statistical survey methods. The group recommended an "MBA-type" cost administrator as a project manager's counterpart and the use of risk analysis and decision theory on all major programs.

At the same time, middle management at NASA began to question the Agency's success-at-all-costs practice giving way to the new low cost initiative. For example, W. G. "Bill" Stroud, Chief of the Advanced Plans Office at GSFC, sent Del Tischler a "question for top management." He wrote:

A point that I would like to raise concerns the human behavioral aspects associated with modifying the Agency's criteria for successful performance. The ground rules laid down in Low's remarks—i.e., a) don't reinvent the wheel; b) standardize; c) design to minimize testing and paperwork; and d) trade performance for cost—will, if pursued, probably make NASA a tremendously different organization. What will happen to an organization which has been proud of its tradition of working at the cutting edge of high technology when that challenge is no longer a primary objective? What are the incentives for a Project Manager or a Center Director to significantly reduce the costs of a payload? Will he get more projects or programs? Will some part of the savings be available to him for his favorite research projects at the Center? Will there be penalties for taking more risks, having more failures?

I suggest that there is a very challenging set of behavioral issues involved in the changes in direction that George Low has indicated, and I wonder to what extent the Agency's top managers are prepared to deal with them.[2]

By January 1973, NASA's low cost team reported to the Associate Administrators' Steering Group for Cost Evaluation. Low had formed this group, made up of the deputy administrator and the five program associate administrators, in the summer of 1972. In forming the steering group, Low was responding to a number of suggestions about the absolute requirement of getting top NASA management behind the low cost drive. Tischler's Special Panel for Space Cost Evaluation, consisting of representatives of Headquarters offices and Field Centers, reported to Low's steering group. Reporting to Tischler's special panel were working panels for Stabilization, Control, Guidance and Navigation; Communications and Data Processing; Power; Attitude Control and Propulsion; Systems Definition; Historical Costs; and Cost/Risk Assessment. In turn, a number of working panels had working groups reporting to them. For example, reporting to Attitude Control and Propulsion were the Inert Propulsion Working Group, the Chemical Propulsion Working Group, the Electric Propulsion Working Group, and the Requirements Working Group.

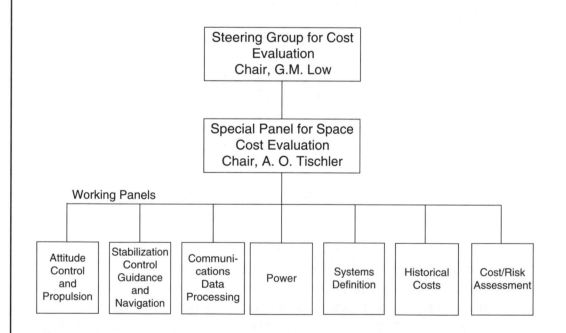

In early January 1973 Tischler sent Low a progress report. It contained good news and bad news, but the bad news dominated. Many of the panels had made little progress, and Tischler observed: "The bulk of NASA development personnel are still pretty reluctant to do anything to change our way of doing business." Tischler then moved to a set of managerial issues he felt needed to be addressed. He categorized them as motivation, education, implementation and enforcement.

Motivation

Tischler thought that low cost program management was such a major departure from the currently accepted management objective that to be successful it would need a new approach. Money would not be helpful; most NASA program managers were close to the pay ceiling. Another option, often used in a bureaucracy to award superior performance, was to increase staff and budget—but this would hardly be low cost. Tischler thought there was not much that could be done for NASA managers. However, he believed NASA should share cost savings with its contractors. He cited the U.S. Air Force experience in the lightweight fighter and the AX Close Support Aircraft programs where "contractors have not only been able to respond but have been highly motivated to better their own expectations and promises. I strongly favor target cost procurements with very strong cost-sharing incentives."

Tischler then addressed the inertia inherent in moving any large organization in a new direction. He told Low of a scheme being used at the Oak Ridge Atomic Energy plant to get that organization to move itself. A series of brainstorming sessions to identify the future of the plant had been conducted. Not only were employees involved in planning their future, they were retooling themselves by attending education classes taught on site, much like TQM (total quality management) training two decades later. Instead of resisting change, the Oak Ridge employees were looking forward to it.

Tischler recommended that NASA study the Oak Ridge scheme, but it did not. He felt a positive attitude toward change was necessary to lower NASA program costs. He concluded: "I would now add my personal comment that motivation is poorly worked in NASA. The excitement and glamour of working to reach and explore space has been adequate for our people. But the cost problem lies at right angles to that attraction and that must be taken into account."

Education

Tischler reminded Low that NASA program managers were rewarded for performance, not schedule or cost. All project managers of space or aeronautical programs were "re-cast" engineers. As such, they obtained satisfaction and rewards for meeting or exceeding performance goals and improving systems beyond minimum requirements. They also tended to be very responsive to technical teams who were constantly attempting to improve their hardware.

To overcome those "built-in barriers to low cost," Tischler recommended "educating" the managers in the use of low cost techniques such as tools and technology that promote low cost. The problem was, NASA needed to design such tools. Tischler reiterated NASA's need for "a methodology for assessing necessary programmatic decisions in terms of cost, benefit and risk. The use of such tools will at least develop insights to make a formidable problem somewhat more tractable. It will tend to make managers manage, and that is important to the cost problem." Tischler then unleashed his version of shock therapy.

Carried further, this logic suggests that project managers need not be engineers; engineers should be used to develop techniques, assist in defining requirements and formulating specifications, and to advise the program manager on the technical aspects of his managerial decisions. (Remember now that I am a professional engineer, and I do not regard this as heresy.) The program manager should have training and experience and some authority to exercise control in organization, management, and other fields of administration that will help him do his primary job, which is to maximize the effectiveness of the resources available to him toward achieving his defined goal. Note the contrast with the attitudes of our engineers, who are intent on maximizing the capabilities of the tools (system, components) used in the project.

Since I don't believe that NASA will voluntarily submit to non-engineering management, I say that some educational retraining of our engineering managers is necessary to place the objective of maximizing the effective use of our resources in view, along with the traditional striving for high performance. Training programs for NASA program personnel might be implemented by contracts with either an accredited institution of learning or a first-class management consultant firm. Initially their assignment should be to explore the problems, and to make recommendations to NASA on the steps to be followed.

Implementation

Tischler pointed out that opportunities to impact cost occur at all levels of management, not just in NASA's programs and projects. NASA clearly needed to do a better job in project definition, as pointed out in the historical panel study, and the associate administrators would have to work hard at determining credible targets for project costs. Tischler then made the case for driving decisions to the lowest level practical. Payloads, for instance, could be generated in large numbers if NASA would let other agencies, private enterprise and the academic community provide them within safety limits and other criteria set by NASA. To do this NASA would need a policy that guarantees any credible payload a ride into space within three months of delivery. NASA must agree to accept the payload on a whole system basis, much in the mode used by the CV-990 operation.

According to Tischler, "A single organization not controlled by the program offices should be given the responsibility to develop standard systems and make them available on a timely manner. No manager can be motivated to use standards that are not available when needed." Tischler declined to say where this office should be located, but observed: "Note that the standardization objectives are applicable whether payload definition is dominated by NASA or not. The experience with the CV-990 indicates that most payload people who are interested in data results rather than in developing payloads are anxious to borrow, rent or buy if necessary, anything that will hasten their flight opportunity—a fact that NASA has overlooked."

Enforcement

Tischler believed without enforcement, changes in the way NASA did business would be slight. Enforcement could be a set of subsystem "czars" to whom all project requests for subsystem development would be forwarded for delivery, or a catalog of standard subsystems would be available to the project community from which they could pick and choose. The ultimate solution should be the subject of "an extremely high-level NASA management seminar," he suggested.

Tischler concluded: "Let me add a few more plain words. These are primarily with regard to the housekeeping subsystems. What's wrong with our costs is not mysterious. We've had that about right from the beginning. What needs doing is not that mysterious either. You had it about right in your AIA speech last spring," he told Low. "But doing it requires decision, organization and authority."[3]

However, as the results of the low cost systems effort were to show, the important decisions were never made; the organization was weak and ineffective, and real authority was never delegated.

Tischler's Paper

During this time Tischler also prepared a lengthy article, "Commentary on Space Program Costs," for publication in a national journal, a very good way to get the word to the world that NASA was serious about cost.[4] He forwarded it for review to Low, who took exception with so many of Tischler's statements that he returned the paper with a note that included: "I have read your paper and find that I disagree with so many of your statements that I cannot approve its publication."[5] Low's motivation for rejecting Tischler's observations may have had more to do with the way that he wished the aerospace industry worked rather than the reality of the situation in the early 1970's. Nevertheless, with the clarity of hindsight, Tischler's original observations were very astute, and many of the issues he numerated still plague NASA in the 1990's.

Tischler's paper began with the observation that if NASA could not lower costs, it would not be able to afford the payloads that the Shuttle would be able to transport. Payload costs were running at 60 to 80 percent of total mission costs and were actually still increasing on a cost-per-pound basis. And if we were to continue developing unique hardware for each mission, the costs would severely limit the number of payloads generated.

Tischler thought that the institutional base then in place to support NASA's project activity was too large for the total project budget, which caused many in NASA and the industry to promote projects in order to keep the support activities busy. These support activities had grown accustomed to a management style identified by the heavy use of support contractors, premium standards of excellence

---

EIA Predicts Erosion of Space Agency's Buying Power

"Another long-term issue will be the continued reduction in the size of NASA's civil service and contractor staff. The EIA study suggests that an effort may be made to redefine the roles of NASA's centers and even close one or more of them."

Space News—October 18-24, 1993, p. 27.

NASA Choices and Challenges

The difference between NASA's earlier and current cost estimates places it in an uncomfortable position. Observers may well wonder why the agency did not fully inform the public about the station's cost. Also, congressional decision-makers formed their initial opinions about the space station based on the earlier, lower estimates; had more complete figures been presented, the project might not have earned the same support. Ultimately, a full disclosure of the estimated cost of the station may prove to be the project's greatest liability.

The GAO Journal—
Number 14,
Winter 1991/92.

demanded by the "no failure" dictate, detailed and comprehensive testing, redundant checkouts, and emphasis on documentation and traceability. All these drove costs up. As Tischler put it, the choice is simple, if brutal; either do it cheaper or give up some of the work.

Tischler then discussed the aerospace culture. The principal characteristics of this culture, he believed, consisted of solving intricate problems and developing perfectly engineered unique systems. The quest for engineering perfection was reflected in overly stringent requirements, specifications and engineering change orders with an accompanying increase in costs. Another important characteristic of the culture was an absolute commitment not to fail. Where the safety of the crew is concerned, there can be no tolerance for failure, but to maintain the "no failure" culture on noncritical unmanned experiments would be foolhardy as to costs. As Tischler put it, "The name of this game should be to maximize winning, not to minimize losing. We must learn to recognize and accept appropriate risks." The culture described by Tischler contained much of the traditional aerospace program organizations and approaches, and as such generated a complex maze of conditions and rules through which decision must proceed. Much of this complexity was intended for large space projects but was often applied to much simpler tasks. The process was stultifying, involving large numbers of people, again with a corresponding direct effect on costs.

Tischler next addressed some cost-contributing peculiarities of the aerospace business. The first of these is the "buy-in," where a company bids a low proposed cost and relies on "getting well" on change orders. "Cost reduction is not a tradition of this business because there is not a penalty or reward to the project manager to reduce costs. Once the budget has been justified and approved for a project, there is simply little motivation to reduce it." As one old timer put it, once the budget is known, the job will never be done for less.

"For the project manager, the emphasis is on performance objectives—the real measure of success or failure, especially among peers. Until working to lower cost is accepted as an honorable thing to do, cost is not a significant parameter. The aerospace business had long been identified as one in which failure must be avoided at all costs. As projects proceed, conservatism sets in and hardware and tests are backed up by more hardware and tests." Tischler told the story of Frank McGee, the television commentator reporting during the Apollo missions. McGee said that with all the redundancy in the program, he would not be surprised if NASA had two moons. All of this contributed to the upward cost spiral.

Tischler noted that the lack of sufficient initial project definition had been identified by a number of groups as a major cost driver. The results of this are seen in the design changes, production halts and schedule delays that occur until firm project requirements are finally identified. This process of redefining requirements and specifications provides the means for the contractor to correct any initial underbid. Tischler explained that some lack of definition can be expected in projects running close to the edge of technology advancement, but much of the cost escalation is caused by the culture.

As to the use of new technology, each project manager should assess methods of achieving objectives with particular attention to conducting tradeoffs of the cost drivers. These assessments should include an evaluation of the maturity of the technology, since the risk of cost escalation is obviously higher when new technology is developed concurrently with the design of a system. To lower costs, Tischler recommended unproven technology development be conducted outside the constraints of a project schedule. In summary, Tischler recommended a clean separation of the phases in the development of a project with renewed emphasis on the definition phase.

The next topic was cost estimation. Since cost was a distant goal in relation to performance, there was little incentive to improve NASA's ability to perform good cost analysis and tradeoffs. He noted: "Cost estimating must be recognized as a highly specialized technical profession on a par with any engineering discipline, and, as such, needs talented, well-educated people, tools and development opportunities. If we are to reduce costs, we need the capability to identify, analyze and understand the things that drive program costs." Tischler defended reliability and quality assurance as being much maligned and frequently offered as a scapegoat for cost increases. "The key here is the project manager's need to define the risks and how much of safety, reliability and quality assurance support is required."

Tischler then looked at different ways of doing business and pointed out COMSAT, a commercial satellite communications system organized to provide service to its subscribers, as an interesting, "different" model in that its satellites were procured to performance requirements only. Details of design and production were left entirely to the selected contractor. All satellite hardware contracts were fixed price-incentive fee with the incentives based on performance and life span.

"This approach," said Tischler, "is predicated on the idea that results rather than method pay off—by clear definition, by conservative application of advanced technology, by use of motivated contract teams, and with a minimum of interference, COMSAT has achieved an enviable cost effectiveness."

Finally, Tischler addressed payloads as an appropriate area for concentration because payloads would be the principal cost element in the era of the Shuttle. The average space mission can be broken down into three main elements: the launch vehicle, the spacecraft or the payload carrier; and the payload, consisting of the specialized science or experimental measurement equipment. Tischler felt the use of the Shuttle, coupled with the standardization of spacecraft subsystems, would save hundreds of millions of dollars a year.

Tischler's paper did indeed contain some debatable assumptions and he may have too vigorously condemned past practice, but his ideas were a true expression of the issues and a vision of how NASA should change. However, NASA was not sold on the need or advisability of reducing costs.

The Historical Cost Working Panel

In addition to Tischler's commentary, Low now had access to a number of studies and reports prepared by the task force. The report of the Ad Hoc Historical Cost

Working Panel of November 1, 1973, was of special interest to him. The team collected and analyzed cost data from 25 NASA projects, including manned and unmanned, Earth orbital and planetary, large and small, with high and low cost growths. The analysis identified high cost areas, areas for improvement and things that were managed well. The group interviewed a number of NASA contractors. They believed a major stumbling block to an even more thorough analysis was the lack of NASA historical data.

Conclusions of the Ad Hoc Historical Cost Working Panel

1. The smaller the number of government people involved, the smaller the size of the contractor organization.

2. The lack of sufficient initial project definition has been one of the key significant cost growth factors. For low-cost programs, it is essential that the evaluation of alternatives be considered an important element of project definition.

3. Technology advancement should be undertaken primarily outside of on-going projects.

4. Experiments and payloads for spacecraft should be selected early in the project definition phase. Final payloads are too often radically different from the assumed payloads used in early definition.

5. Contractors feel that substantial cost savings could be realized if the government would shift the emphasis from detailed monitoring of Phase C and D development to better definition in Phase A and B requirements.

6. There needs to be an improvement in cost estimating methodology. NASA does not have common cost estimating guidelines, work breakdown structures and terminology.

7. The tendency of many program managers is to blame high reliability numbers and stringent quality control specifications for many of their problems. Most program costs do not justify this claim. Reported costs average only six percent of total program cost.

8. The cost growth impact of schedule changes on manned programs has been more severe than in unmanned programs. It is cheaper to slip programs early in the development phase and get better definition before high manpower loadings start.

9. Although reportable documentation costs run up to four percent of total contract costs, the true costs have been estimated to be as high as

15 percent of total program costs. Rapid documentation approval would reduce costs.

10. The contractors responded by two to one that NASA specifications were too stringent.

11. Comparing the distribution of final costs between projects is dependent on the work breakdown structure used by each project. If management wants visibility into projects, it should impose a common structure.

12. Analyses of cost-to-complete data for over 20 projects showed that about half had growth factors above 2.0.

Recommendations:

1. Improve NASA's cost estimating capabilities by establishing the systematic collection and storage of cost and technical data. A focal point must be assigned the responsibility for establishing such a data bank.

2. Develop training programs for engineering-estimators. Institute rotation of personnel both inter-Center and intra-Center. Impress on our current program directors that it is equally important to untrain those project managers raised during the era of plenty.

3. Expand the ad hoc operation of this study activity and elevate it to permanent organizational entity reporting at an appropriate management level.

4. Senior management should provide rewards for adhering to cost controls just as they have for successful technical performance.

5. Finally, if money is saved by dedicated cost reduction programs, that at least some of the savings should be used to increase the technology efforts in the Agency.[6]

Armed with all these studies and reports, Low initiated a permanent office in April 1973 with the appointment of Del Tischler as Director of the newly formed Low Cost Systems Office (LCSO). In September 1973, NASA Management Instruction #129.2 was issued to formally define the roles and responsibilities of the Director of the Low Cost Systems Office. The Director was "to lead a program to introduce specific means for reducing the cost of NASA Space Systems," and his duties were to:

1. Conduct studies of factors affecting cost of space programs; based on these studies, recommend a program to reduce the cost of space missions.

2. Identify and evaluate specific business approaches and management methods which will reduce the cost of NASA programs; assist program offices in implementing low cost techniques.

3. In coordination with program offices, lead a program to develop standard equipment to be used for NASA space programs.

4. Coordinate with the Department of Defense and other government agencies to establish standardized components and subsystems.[7]

Low Cost Systems Office Organization

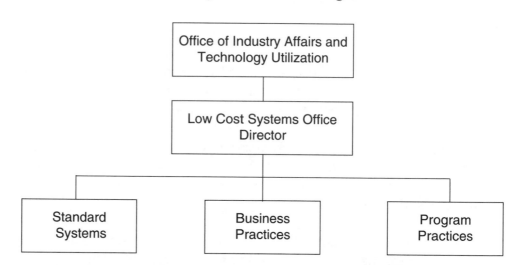

Standard systems would be the backbone of the LCSO activities. Standards would be developed in five areas: electric power; communications and data handling; auxiliary propulsion; stabilization, guidance and control; and ground support equipment.

The business practices group would review and evaluate NASA's planning, project approvals, scheduling, staffing levels, procurement, contract execution, and sustaining activities with a view to instituting new business methods that were expeditious and cost-effective.

The program practices group was to survey, review and evaluate program factors which significantly affect the cost of space program development, production, operations and support to identify opportunities for effecting future savings.

However, Low's decision to have the newly established office report to the head of Industry Affairs and Technology Utilization rather than himself sent an unwanted message to the NASA leadership, especially to those at the Field Centers whose support and participation were critical to the success of the effort. The message was that cost reduction was not so important after all since Low did not have the new office report to himself or to an important direct report office. The Office of Industry Affairs and Technology Utilization was not considered by many in NASA to be powerful and influential enough to bring about necessary drastic changes in NASA's culture. To the tea leaf readers, this organization arrangement did not portend well for the future of the LCSO.

Footnotes

[1] A. O. Tischler, personal correspondence to author, August 10, 1994.

[2] W. G. Bill Stroud, "Cost Improvements and the Shuttle," memorandum to A. O. Tischler, August 9, 1972.

[3] A. O. Tischler, memorandum to George M. Low. Washington, D.C., NASA, January 4, 1973.

[4] A. O. Tischler, "Commentary on Space Program Costs," unpublished manuscript. Washington, D.C., NASA, February 1973.

[5] George M. Low, memorandum to Del Tischler. Washington, D.C., NASA, February 20, 1973.

[6] Historical Cost Panel, "Report to the NASA Space Cost Evaluation Program," Washington, D.C., NASA, November 1, 1973.

[7] NASA NMI 129.2, September 1973.

Chapter 5

Standard Equipment:
The Backbone of the Effort

When George Low launched NASA's campaign to reduce cost, he thought mostly in terms of hardware. Tischler described it well in a paper given in July 1973 to the combined meeting of the American Institute of Aeronautics and Astronautics (AIAA), the American Society of Mechanical Engineers (ASME) and the Society of Automotive Engineers (SAE) in Denver, Colorado.[1] "Standard equipment is the backbone of the total effort because the space program is populated by engineers and scientists, and whether or not they are conversant with management, programmatic or procurement practices, they do understand hardware." This proclivity, combined with NASA's tape recorder experience and the payload potential of the Shuttle, led the Low Cost Systems Office management and staff to concentrate on developing and cataloging standard equipment. After all, experts were still warning that the real problem at the outset of the Shuttle era would be a lack of payloads to fill up the Shuttle's cavernous cargo bay. The use of standard and available equipment would, in theory, allow payloads to be flown in the Shuttle quickly and relatively inexpensively, thus fully utilizing the Shuttle capabilities and appreciably lowering payload costs.

There was a sense of urgency in getting on with the hardware portion of the job. In June 1973, Tischler circulated for approval a proposed NASA Management Instruction (NMI) covering the Standard Space Systems Program.[2] It contained the basic policy governing the development, acquisition and use of standard equipment. More than one year later, in October 1974, just a few months after Tischler departed NASA, the NMI was finally approved. The intense interest in this directive was evident in the amount of time the NMI spent in the coordination and approval cycle. The idea of mandating the use of standard equipment in NASA programs provoked the NMI reviewers to howl with indignation: "ridiculous," "could never work," "the Centers will torpedo this," "it will require *more* NASA and contractor people, not fewer," "the accountability and responsibility is poor, with everyone able to blame other Centers for schedule problems," and "NASA is just too big and scattered an organization to try to apply the proposed type of matrix organization."

Why was almost all of the NASA leadership against this directive? The major reason was the issue of control. No level of NASA management, from the

Headquarters program associate administrators to the Field Center project manager, was willing to see their control diluted in any fashion, and the proposed NMI had a remote Headquarters office dictating equipment usage down to a subsystem item, constituting an enormous loss of control. What would this office do next? Tell the Centers how to manage? Despite these serious reservations on the part of many NASA managers, Low was determined, and the NMI became the law of the land.

Just how determined Low was is reflected in the following memo to Gray (E.Z.) and Tischler (Del).

August 3, 1973

E.Z. and Del—I am not in a compromising mood—and don't believe that I am lacking in guts or conviction. Let the first project manager who does <u>not</u> want to standardize—when standardize he should—come to my office to tell me so. My door is open! —GML (George Michael Low)

Under this controversial new policy, NASA would standardize subsystems and components, thereby spreading costs over several projects and gaining additional savings through quantity procurements, plus increase reliability. Program and project managers would be provided a range of low cost standard subsystem and component options suitable for a majority of future missions. The LCSO would translate common requirements into standard equipment specifications and would also review new programs in conjunction with the Field Centers, to determine the feasibility of using standard equipment. It would identify available equipment designs; justify new designs; propose the design, development and qualification of new standard equipment; and coordinate the development of NASA standard equipment with the designated Field Center.

The LCSO would produce and maintain a catalog of standard subsystems and components. The program offices would retain responsibility and authority for their projects, but were expected to use standard equipment. Disagreements between the program offices and the LCSO would be resolved by the Deputy Administrator. Field Centers would help identify candidate standards, perform technical tradeoffs, conduct cost effectiveness studies and, upon development approval, would design, develop and qualify the standard. The Center would provide the LCSO with project status and information for the catalog to assist potential users. Panels would be formed in support of the LCSO and would be chaired and staffed by Center personnel. They would review, critique and recommend subsystems and components for consideration as standards. The funding for development of standard equipment would be provided by the program office identified as the first user, with the LCSO funding additional costs attributable to developing a standard. The user would pay for standard equipment subsequent to development.

By early 1974, as the Standard Space Systems NMI was being circulated for approval, Tischler decided to set the date of his leaving NASA. From the very beginning of the effort, Tischler knew he was not completely attuned to Low's vision of a low cost NASA, a vision he thought too narrow. As an Apollo program manager, Low understood well the expense of developing specialized equipment designed to exacting mission requirements. Therefore, he embraced fully the idea of "standard" hardware items. Tischler, however, saw standardizing pieces of equipment as only a first step in a much larger program of cost control. To him the entire managerial logic of NASA's program methodology was a target for revision. He frequently said, "Only a limited number of people in NASA do the real work but everybody wants to get into the act." He thought that large numbers of NASA executives drawing considerable salaries could not contribute to technical decisions because they lacked the knowledge or the experience. They seemed to seize loose ends of programs, magnify their significance, form committees, panels and boards to review and critique progress, write and revise regulations and procedures, and hold up programmatic approvals until they could be sufficiently educated about current problems by the program personnel. When they finally formed a consensus among themselves, Tischler thought they imposed their ill-found findings onto the programs without regard for the effects on schedules and costs. The consequences were that everything seemed to take longer and cost more.

Tischler believed many of NASA's cost problems stemmed from this management overpopulation and underproduction. To solve this he would have cut NASA's overhead staff by half. "Which half isn't critical," he said. "The remaining half would get smart quickly. But if I were doing it, I'd start with the Associate and Deputy whatevers and their titled assistants, too; those who only monitor what's going on in order to make viewgraph presentations to successive levels of supervisory authority. It's an extravagance that only a government can afford.

"We have no option but to write different management equations if low cost is to be realized. Whoever does so must be prepared to be vilified, publicly and in concert, by the government Agency's staff, by contractors whose profits depend on Agency programs, and even by members of Congress who draw significant campaign contributions from their contractor constituents. But it will have to be done."[3]

Tischler said such outbursts on his part caused Low much anguish. Low felt that the cost effort would be jeopardized if it antagonized "our own people." It was clear that there were large differences between Low and Tischler on the scope of low cost efforts. Tischler thought Low had hoped that the Low Cost Systems Office would eventually generate a catechism to guide NASA's future program managers; but from the beginning, Tischler was going for reformation.

Tischler did an outstanding job in bringing his and Low's concepts of low cost together, at least enough for an effective start. He believed in the cause, but felt Low had let him down in not acting aggressively enough. It was inevitable that these contrasting views would clash. When confronted with Tischler's management scheme for the low cost program effort, Low equivocated. Low then made what Tischler considered a fateful decision. Low removed Tischler and the newly formed office from reporting to him and placed them under the Office of Industry Affairs and Technology Utilization. That was the straw that broke Tischler's resolve; he decided then and there to retire.

Nov. 3, 1993

Robert Frosch, Testimony to the Senate Committee on Commerce Science and Transportation Subcommittee on Science Technology and Space –Washington, DC.

"We cannot continually substitute bureaucratic trivia ('mickey mouse') for real technical work, and expect anything useful and productive to result. It is easy to blame the 'bureaucrats' in NASA for the nonsense, but they have been driven to their behavior by forces largely outside the Agency, in the Executive and Congress. Over the years, increasingly bizarre processes and even more complicated formal systems have been invented as attempts to solve real and fancied program problems, until, by now, the processes have merely become an inadequate substitute for doing anything properly."

59

In July 1974, in anticipation of the Standard Space Systems NMI being signed, George Muinch, the new director of the LCSO, and members of his staff met with NASA Field Center representatives to initiate the day-to-day procedures for the development of standard equipment at NASA.

Muinch was a solid replacement for Tischler. First and foremost, he started with a clean slate; he was an outsider not imbued with the Apollo culture. Muinch was an entrepreneur and as such was intimately involved in bottom line cost identification and control. He knew the aerospace industry from the industry side, and he was hard nosed. He would not accept "but we've always done it that way" as an excuse for an inability to implement change. On the downside, as an outsider he was suspect by old line NASA management, his NASA membership dues being about twenty years in arrears, and he did not treat NASA obstructionists kindly.

Muinch had one other powerful advantage. He was fully committed to the task at hand, and during his leadership from 1974 to 1977, the Low Cost Systems Office made significant progress in implementing this new management initiative.

George Muinch was born in Indiana. When World War II broke out he joined the Navy as a V-5 Aviation Cadet. In July 1943 at age 20, Muinch was commissioned a Naval Aviator. He was sent to the Pacific theater in March 1944 and assigned to Carrier Air Group 20. This was the air group famous for sinking more enemy tonnage and destroying more enemy aircraft and ground facilities than any other.

During a bombing mission on the harbor and airfield in Taipei on October 12, 1944, Muinch scored a direct hit on an enemy ship but was struck by anti-aircraft fire. His aircraft headed down, but he and his tailgunner elected to stay with it and prepared for a water landing. Just when things couldn't get much worse, Murphy's Law kicked in—a Japanese gunboat suddenly appeared on the horizon. Since the boat's proximity to where Muinch planned to ditch meant almost immediate capture, he emptied his 20 mm wing guns into it, hoping for a

George Muinch and Edward "E. Z." Gray.

Courtesy of George Muinch

disabling hit. Muinch and his gunner survived the water landing and got into their survival raft, remaining in the water from just after dawn until shortly before dusk, listening to the gunboat attempting to start its engine. Finally it started. Just then Muinch was startled to see an American submarine, the USS *Sailfish*, pull alongside his raft. It was a rescue submarine alerted to the ditching by a fellow Air Group pilot. The submarine, using its deck gun, quickly drove off the gunboat and pulled Muinch and his crewman on board. The next day the sub needed to surface to recharge its batteries. The captain asked Muinch to serve as an aircraft spotter. Muinch and four others were on deck when the *Sailfish* was attacked by a Zero. Having been briefed on the frightful consequences of not making it down the hatch on a crash dive, Muinch jumped for it. In the process he broke off the heel of his shoe and sprained an ankle, but he made it.

Shortly after Muinch came on board at NASA he had a visitor from the Kennedy Space Center. The *Sailfish's* executive officer, Pat Murphy, now a member of the Center director's staff, recognized Muinch from the photo in the NASA Activities magazine and stopped by to once again welcome him on board.

Muinch was off to school following the war. He graduated from the University of Notre Dame with a BS in Physics and received a Master's degree in Applied Physics from the University of California, Los Angeles. For the next 30 years Muinch worked mostly in the aerospace industry, specializing in research and development involving electronics and physics application. He worked with the Hughes Aircraft Co., Lockheed Missile Division, Aeronutronic Systems, Inc., Northrop Corp., Dynascience Corp., and the Computer Measurements Corp. Upon retirement in 1972, he moved to west Texas to take up cattle ranching. Muinch's record of success in private industry was outstanding, and it was this record that brought him out of retirement. Low asked George to come to Washington for a visit and promptly offered him the job. Muinch agreed to join NASA but put a time limit of two years on the job. He started work in May 1974 and left in April 1977, one year beyond his agreement.

Muinch took the job because he felt costs could be lowered. From his industry experience he knew this was not a fool's errand. But from the beginning, Muinch believed it could happen only if the NASA Administrator remained firmly behind it. He knew the business from the contractor's side; he knew what was technically feasible and what was practical, and he knew it would not be easy. He took the challenge only because he was confident that with the proper support, he could succeed in reducing costs.

Standard Equipment

The standard equipment program posed an interesting management dilemma for Muinch and the LCSO staff. Identifying and developing standards could not be done by the LCSO alone. In fact there was little of the job that could be done by the Headquarters office. Bringing standards into being had to be accomplished by and at the NASA Field Centers. Without their help the job simply could not be done.

One way of involving the Centers, much used by NASA in the past, was to form multi-Center working groups, panels or other forms of special teams combining Center technical and managerial expertise to accomplish Agencywide goals and objectives. The Center identified to develop the standard would, of course, be very much involved, but there was much more to the total job. For both practical and political reasons, much of the work required a multi-Center effort.

Tischler had included the use of panels in the proposed NMI. Muinch agreed to this approach and added his thoughts to the list of panel duties. First Muinch addressed the role of the panelist. He wanted panel members to be technically competent and able managers. They had to have authority to speak for their Centers on standard equipment issues; they had to have access to their management and other Center experts to make technical, cost and programmatic comparisons of candidate standards. They would meet with the DoD, government agencies, industry and other potential users to market NASA's standards and they would advise on the business and program practices effort.

Muinch wanted the panels to be responsible for the full standard development cycle from identifying candidates for standardization to equipment development and logistics.[4] Panels would develop the technical requirements, prepare lists of candidates, establish performance and reliability specifications, and develop selection guidelines. Then they would support the cataloguing efforts. To be successful, the panels had to develop standard equipment responsive to the users' requirements at a minimum cost and in a timely manner.

As to the selection criteria to be used for standard equipment, Muinch proposed three major groupings: technical, cost and schedule. Standard equipment designs could take several paths:

- Use of current, flight proven equipment

- Modification of currently proven equipment

- Development of equipment from previously proven technology

- Development of equipment in a full-scale, multiphase program

Standard equipment should result in a significant savings for NASA and be proportional to the cost and length of time needed to produce the standard equipment, Muinch thought. Finally standard equipment development should take anywhere from 1 to 12 months and in some cases as long as 18 months.

Muinch used the following chart to describe the flow of the standard equipment process. The chart was deceptively simple. In reality the process was full of mine fields that were triggered by the slightest perturbation.

Recommendations for standard equipment development could come from a variety of sources: NASA Centers and projects, the Department of Defense and other government agencies, especially those that are historical users of space and the commercial world such as COMSAT. Once a candidate was suggested, a pre-

The Standard Equipment Process

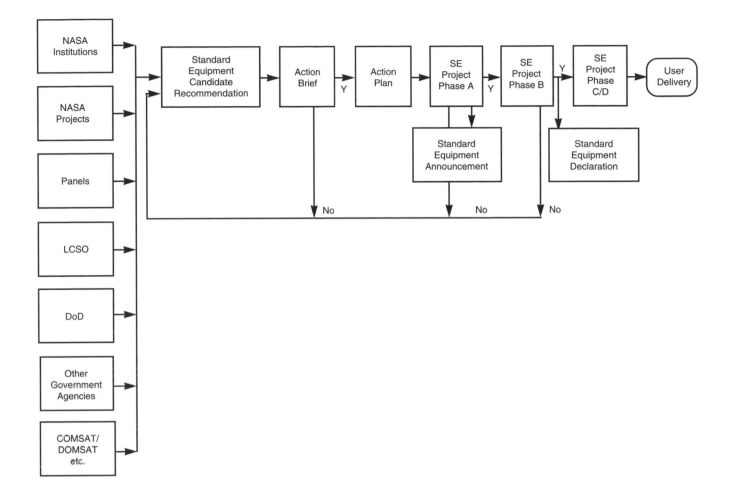

liminary assessment would be made, including the schedule and quantity of the equipment needed, along with a first cut cost and saving assessment. If a positive assessment were made, the panels and the LCSO would prepare an "action brief."

The identification and selection of the standard equipment and the first user (and therefore developers) of the standard were fraught with danger. No one wanted the job, and the panel members would frequently be less than candid in attempting to ensure their Center did not get stuck with the development task. Part of this aversion was the manner in which the NASA culture allowed its project managers almost full discretion in exercising their management prerogatives.

Dr. Robert Cooper, a former director of the GSFC and as such intimately involved in the development of standards, could even recall managers who attempted to tinker with launch vehicles to better accommodate their projects. Everyone knew the managers would not welcome the additional responsibility of developing a standard. For instance, at a meeting to finalize the development of NASA's first standard tape recorder, the GSFC project manager of the identified first user project was called in and given the good news. His response was worthy of a Hollywood screen test. He virtually exploded. He began by questioning the parentage of all those present. Next he swore he would never use equipment he had not developed specifically for his project, and then he stormed out of the room. When faced with a personal confrontation with the Deputy Administrator of NASA, an event that could be very career limiting, he relented, and NASA was on its way to developing its first piece of standard equipment.

The next two steps of the process, the preparation of an action brief and, if approved, an action plan, involved the essential paperwork needed to start the standardization process. The action brief was a proposal and contained only enough information to allow for a go or no-go decision. If it were a go decision, the action plan was prepared. The plan contained enough information to fully proceed with the development. These apparently reasonable and logical steps violated

---

Action Plan Suggested Outline

1.0　Introduction
　　1.1　General description of equipment
　　1.2　Users—firm and potential
　　　　　Describe type/class of S/C use why not used in other types of S/C
　　1.3　Previous history (why aren't previous equipment ok for future S/C?)

2.0　Equipment Description
　　2.1　Functional
　　2.2　Design

3.0　Project Description
　　3.1　Project history including investment to date, and current status
　　3.2　Project objectives

4.0　Work Plan

5.0　Schedule (related back to work plan)

6.0　Management Plan (LCSO is not part of project management but is in a
　　　　coordination and support role)
　　6.1　Current project
　　6.2　Engineering maintenance and modification
　　6.3　User support-candidates

7.0　Procurement Plan

8.0　Documentation Plan
　　8.1　Drawing spec tree, quality progress, provide for future projects, etc.

9.0　Resources
　　9.1　Dollars, manpower (civil service, contractor), facilities

---

a major NASA tradition, for they introduced new paperwork requirements to the NASA system. Both the action brief and the action plan required new report formats. In any top ten lists of complaints about the low cost effort, the paperwork issue would always be included, the complaint centering on why the LCSO did not use existing NASA processes and paperwork.

The next step concerned the availability of project managers, a common problem for a NASA project managed by a Field Center. Good project managers are never easy to come by. Developing such talent usually takes many years of training, usually on the job, along with careful mentoring and job rotation. With the influx of standard development projects, many Centers were absolutely swamped and had difficulty providing qualified managers.

When the standard equipment development was finally ready, usually signified by a negotiated development contract, the standard was declared official and had to be used on NASA projects unless a waiver was obtained.

The process appeared simple but proved difficult to administer, as witnessed by the changes the head of the Office of Industry Affairs and Technology Utilization, Ed Gray, approved in early 1975. Gray wrote to the NASA Center Directors that many questions had been raised regarding the organizational responsibilities involved in the selection, development and furnishing of standard equipment. Problems in the operations of the panels were also bothering Gray. The panels were not yet fully staffed, there was no extra travel money to attend panel meetings, the panel representatives frequently were not empowered to represent their Centers, and there were numerous complaints of "cumbersome procedures." After two years of experience developing only ten standard components, Gray recognized that changes were necessary. New procedures were needed to smooth the standard equipment process, at least in the short term. Gray proposed the following:

• The selection and prioritizing of candidate standard equipment remain the function of the Low Cost Systems Office and the NASA Field Centers. However, recommendations can also originate from Headquarters program offices.

• The Low Cost Systems Office will notify each program office and Field Center of the candidates for standardization and ask for their recommendations for or against. If the recommendation is positive, performance requirements and potential first users will be identified. A date will be set for a program plan review, at which involved Centers and program offices will be invited.

• At the program plan review, a determination will be made to proceed or not with the equipment under consideration. If the decision is to proceed, a "first user" project will be identified and the program plan adopted.

- The "first user" project will normally be designated the developer of the standard. They will manage the standard development including exercising specification control. The managing Center of the "first user" project may wish to establish a separate organization to provide the standard equipment as Government Furnished Equipment (GFE) to all future NASA users during its lifetime as a NASA standard.

- As a rule, the most reasonable and cost-effective approach to supplying users' standard equipment will be GFE. When the GFE is not used, a justification would be submitted to the Low Cost Systems Office.[5]

Gray reminded NASA management that standardization required NASA organizations to work together and that the "carrot" would be additional funding for organizations who took the lead in specific equipment areas, a promise he never fulfilled.

Was the standardization effort successful? From just about any objective criteria, the standards program was very successful. Three pieces of the originally developed 22 standard components—the transponder, tape recorder and inertial reference unit—are still flying. A detailed cost analysis of the savings to NASA from just these three would be impractical, but the savings must be in the tens of millions of dollars. The main problems were the substantial resources and management attention devoted to it at the expense of the potentially more profitable practices effort.

The Catalog of Available and Standard Hardware (CASH)

The NMI establishing the standard equipment program also provided that "a catalog describing the standard systems or components will be maintained by NASA." Hardware cataloging, however, had started earlier, in June 1973, when JSC contracted with Thompson, Ramo and Woolridge (TRW) to develop a catalog of Candidate Shuttle Payload Instrumentation. This catalog consisted of Skylab and Apollo sensors and experiment hardware, including instruments, flight support subsystems and the accompanying ground support equipment. As mentioned previously, owing to budgetary limitations and the historically high costs of payloads, NASA management was very concerned that the Agency would not be able to afford the number of payloads the Shuttle could accommodate. If the payload mission model used to justify the Shuttle had any basis in fact, payloads would indeed be the limiting factor in Shuttle utilization. This catalog served to identify experimental Skylab and Apollo hardware for possible use in building inexpensive Shuttle payloads. TRW was tasked to develop a database, to inform potential users of hardware characteristics, and provide information to NASA's Equipment Visibility System, the inhouse system used to identify available equipment items by their source, condition and location. Potential users of this information would include NASA, other government organizations, the aerospace community, principal investigators and universities.

The work was successful, and in December 1973 JSC contracted for a follow-on effort which concentrated on additions to the catalog, automating the catalog data system, refining criteria for hardware selection and adding a library documentation system.

The next cataloging effort focused on satellite housekeeping systems such as electrical, power, stabilization and control, auxiliary propulsion, and communication and data handling. The Aerospace Corporation was selected to prepare this catalog.

Existing space hardware, including military satellites, was inventoried next. Upon completion of this inventory, NASA proposed all new start projects be reviewed against lists of hardware having potential transfer.

With these efforts underway, the LCSO negotiated with MSFC for the preparation of an all-encompassing catalog to be known as the Catalog of Available and Standard Hardware (CASH). This catalog was intended to document the systems and components declared standard by the LCSO, along with flight-qualified and

Status-Change Sheet

Cash Cat. No.	Standard Component	Revision No.	Date
1.001	Standard Fixed Head Star, Tracker (FHST)	Original	4/22/75
1.002	Standard Dry Gyro Inertial Reference Unit (DRIRU)	1	10/29/76
1.003	Standard Reaction Wheel (SRW)	Original	11/2/75
1.004	Standard Mod II Fixed Head Star Tracker (MOD II-FHST)	Original	12/31/76
1.005	Standard High Performance Inertial Reference Unit (DRIRU-II)	Original	12/31/76
1.006	Standard Fine Pointing Sun Sensor (SFPSS)	Original	4/29/77
1.007	Standard Precision Sun Sensor (SPSS)	Original	4/29/77
2.001	Standard Hydrazine Attitude Control Thruster Valve Assembly (T/VA)	1	12/8/75
2.002	Standard Propellant (Hydrazine) Control Assembly (PCA)	1	1/2/76
2.003	Standard Initiator	3	8/30/76
2.004-2.005	Standard Pyrogen Igniter	1	3/18/77
2.006	Standard 5-Newton Hydrazine Thruster	Original	4/29/77
3.001	Standard Specification for Silicon Solar Cells and Cell Covers	1	12/10/74

Typical contents of the CASH Catalog as of June 1977.

flight-proven hardware still available to potential users. Standard equipment was defined as those components required to be used on all future programs unless an exception was granted by the LCSO Director. Available equipment included subsystems or components designed, developed and proven from previous flights and recommended for use in current programs. This catalogued equipment had to be considered for use before new procurements were made.

In early 1977, members of the LCSO met with Ed Gray to present the results of several months of evaluation of the CASH system by LCSO staff, industry and NASA users. NASA also hired a contractor to aid in the evaluation. This group agreed that the "available hardware" portion of the CASH Catalog was totally ineffective and would be very costly to turn into a useful tool. It was too difficult to track and account for the equipment in a timely manner. The attendees agreed to halt all CASH-related computer activity at the MSFC. CASH would continue, but only the "standard hardware" portion would be maintained. Another look at the information in the standard equipment announcements would be undertaken, along with the possibility of the NASA Scientific and Technical Information Office taking over the control, printing and distribution of the CASH Catalog. The new catalog of standard hardware retained the CASH title, but was to be reformatted.

There were two major problems associated with the cataloguing efforts of the LCSO. One was the resistance of the Headquarters procurement organization to assist the effort. The procurement personnel constantly criticized the lack of competition in standard equipment procurements and wrote many memos on the subject, mostly for the protection of the sender rather than as assistance to the addressee. CASH did not incorporate any re-procurement specifications, which implied to the procurement staff continuous sole source procurements—an anathema to those tasked to ensure full competition among NASA suppliers.

Another problem was the amount of resources necessary to continually update and maintain the catalog. It was a complex, difficult, time-consuming job that no one really wanted.

Logistics Problems

One of the most difficult of all obstacles to the development and use of standard equipment was the continued timely supply of an item. Conventional wisdom stated that if fully qualified and reasonably priced flight hardware was readily available (preferably off-the-shelf), project managers would queue up to use it. The longevity of the standard transponder and tape recorder, both developed in the early 1970s and still in demand today, seems to bear this out. However, the added responsibilities associated with developing an Agency standard was simply not acceptable to Center managers. First, such a development was extra work for the Centers. They had to accommodate specifications, requirements, documentation, testing and modifications over and above their immediate needs. They had to meet schedules other than their own, and the development had to be successful. Second, all this extra work added few if any extra people to their work force. The low cost initiative came at a time when the Agency was locked into a constant-level budget from the OMB, when inflation was in double digits and when personnel staffing levels were generally being reduced. Standard equipment meant more work for very little or no extra funding, and while Low thought it was good for the Agency, there was nothing in it for the Centers. If the development failed or if they missed schedules, NASA customers would complain loudly.

At GSFC, where a large number of standard developments were underway, managers worried that they were in danger of becoming the *de facto* NASA Standard Hardware Logistics Center, a role they shunned. GSFC anticipated that by FY1977, more than 35 work-years of effort would be devoted to LCSO work, with the possibility of adding 10 to 25 more, depending on the number of new standard equipment customers. The controversy over resources needed to support the LCSO work continued until the very end of the program.

As originally envisioned by LCSO management, a Center "first-user" project would be identified, and that Center and Headquarters program office would be responsible for the development and logistics of the standard equipment for its lifetime. The equipment would be provided to a customer or the customer's contractor as Government Furnished Property (GFP). Another option, that of Government Specified Equipment (GSE), allowed the contractor building the standard equipment to provide the equipment directly to the spacecraft development contractor. Each option had its avid, vocal following within NASA. In spite of all these problems the use of standard equipment prevailed by virtue of economic reality.

PRC Study

In order to determine once and for all which logistic option was best for NASA, a study was awarded to the Planning Research Corporation. The study was to determine the relative merits and advantages of the two principal techniques for acquiring and supplying standard equipment—GFP and GSE. The study concentrated on the option selected by DoD, the factors influencing the selection and its advantages. The study, "Logistics Plan for the Standard Space Systems Program," was completed in December 1976.

Several factors influenced DoD to select one or the other of the options. They included technical considerations, costs, schedules, quality and standardization. Technical considerations involved complexity, firmness of design/specification, suitability and availability, effect on the development contractor's system and performance.

It had been assumed that GFP was the most cost-effective approach, but the study could not find sufficient data to confirm or deny this assumption. Numerous costs are associated with GFP in maintaining accounting, engineering and logistics control systems. The study did find that the major reason for using GFP (other than cost) was to facilitate standardization.

The results of interviews with DoD personnel revealed that:

• GFP is the most desired procurement technique.

The Barren Shelf

Wesley T. Huntress Jr., Chief of Solar System Exploration for NASA, is quoted in the Feb. 15, 1993, Aviation Week & Space Technology that the agency has saved millions if not billions by using off-the-shelf hardware and software. "The trouble is," he said, "the shelf is soon bare." There's not enough high technology out there to meet NASA's increasingly complex needs.

- Standardization is enhanced by the use of GFP.

- There is no difference or advantage to either approach in terms of government responsibility for performance.

- GSE is not practical for the NASA standard equipment program because of the need to order in quantity without incurring additional production start-up costs, the need to place total quantities through a single procurement source to ensure schedules, and the need to avoid potential systems contractor conflicts.[6]

Multi-Mission Spacecraft (MMS)

The concept of providing NASA, other government organizations and ideally, commercial operators, with a Low Cost Multi-Mission Modular Spacecraft had long been a dream of GSFC employees Joseph Purcell and Frank Cepollina. They reasoned that in the Shuttle era, such a spacecraft would be even more valuable. Designed with serviceability and maintainability in mind, the spacecraft would be brought into the Shuttle bay where housekeeping modules could be repaired or replaced or experiments changed out. A spacecraft built to handle a wide variety of missions within a specific class—and one that could be repaired and maintained in orbit—would be the ultimate in standard equipment.

In developing a concept for standard spacecraft, several considerations had to be addressed. It had to be usable for a variety of projects, meaning that its design would be adequate instead of optimum for each project. Each project would have to be tailored to the standard spacecraft capabilities, but the standard spacecraft would be adaptable to project-specific modifications. Procurements would depend on the availability of block-buy funding. Multiple Centers would now be involved in resolving issues of a spacecraft being built separate from payloads, which would require clean interfaces and effective communications. To be successful, a standard spacecraft program had to avoid repetitive, nonrecurring costs, achieve economy of numbers on recurring costs, increase confidence in cost and schedules, facilitate payload development independent of the standard spacecraft, and use common spares and ground support equipment.

Purcell and Cepollina began marketing the MMS concept in 1972. With the advent of the Low Cost Systems Office, they had a strong potential ally. During his visit to GSFC on February 12, 1973, Purcell briefed Low on the MMS concept, especially the interface requirements such a spacecraft would have with the Shuttle. Despite Purcell and Cepollina's enthusiasm, it was not until November 1974 that the Low Cost Systems Office was instructed to conduct a "Standard Spacecraft Bus Assessment." The general guidelines for the conduct of the study stipulated that three classes of buses would be explored to include missions launched from 1977 through 1984. All three were usable with both the Shuttle and expendable launch vehicle.

1. Scout-Class Spacecraft Buses— Heat Capacity Mapping Mission (HCMM) and Stratospheric Aerosol and Gas Experiment (SAGE) missions as potential first users of a standard bus

2. Large Spacecraft Buses—Television and Infrared Observation Satellite (TIROS)-N, Sea Satellite (SEASAT) A and Large Scale Telescope (LST) missions as potential first users

3. Planetary Spacecraft Buses—Pioneer Venus and Mariner Jupiter-Saturn as potential first users

The bus assessment would include seven current spacecraft and evaluate the following:

- Characteristics and description: structure, communications and data handling, power, attitude control, auxiliary propulsion and unique capabilities

- Technical assessment of candidate spacecraft

- Modifications required to perform each mission

- Technical assessment on integration of payload to bus candidates

- Weight estimates by mission for spacecraft bus, instruments and integration

A cost estimate also would be developed for each candidate spacecraft.

The assessment was conducted under the direction of George Muinch. The study team included the Comptroller's Office, all mission program offices and all Field Centers. Ten spacecraft system contractors provided data for the study.

The GSFC's Multi-Mission Modular Spacecraft Bus was selected as the NASA standard for large Earth orbital missions. The rationale for this selection was based primarily on the benefits of modularity.

- Modularity provides the greatest flexibility to accommodate mission requirements and, therefore, the greatest cost savings.

- Modularity provides the greatest potential for Shuttle on-orbit servicing.

- Modularity provides greatest use of standardized equipment and industrial participation.

- The majority of missions are GFSC; therefore, the interfaces among Centers are minimized.

As approved, the MMS was a 1,300-pound spacecraft, modular at the subsystem level, using standard components. Cost savings were estimated to be over $20 million per project, based on a 13-flight, 11-project comparison. GSFC was assigned responsibility to develop the MMS and provide the buses and operational support services to all users.

The MMS approval document clearly stated that all NASA missions in the category would use the MMS unless waivers were obtained. It was stipulated that other government agencies, foreign governments and commercial customers could use the spacecraft.

In early 1976, the NASA Associate Administrator approved the development of the Solar Maximum Mission as part of the FY1977 budget. This project would be the initiation of the NASA standard modular spacecraft bus.

As it turned out, Solar Max proved the wisdom of the original standard equipment effort. Even when the fine pointing instrument failed, the spacecraft's modu-

Solar Max Mission.   *Photo courtesy of NASA*

lar design enabled an on-orbit repair while other components were returned and refurbished for the next flight. Modularity made the difference.

The MMS turned out to be the single most significant user and developer of standard equipment. With the LCSO, the MMS project worked to incorporate the JPL developed standard transponder and the NASA standard high performance inertial reference unit in their spacecraft. The funding was exactly as planned, the user fully funding the first flight units and the LCSO covering all other nonrecurring costs associated with developing standards.

What made this approach work well had more to do with the actual cost savings than anything else. The MMS project wanted to deliver the highest performing spacecraft for a fixed budget. NASA was in a period of financial constraints and customers understood the value and necessity of using previously developed or standard hardware in the building of a low cost spacecraft.

From 1976 to 1992 NASA built and flew seven separate MMS based missions. The value of each payload was over $150 million with one costing $700 million. Yet in all the missions, the cost of the spacecraft buses was well under $100 million. This outstanding cost performance was directly due to the use of standard and previously developed hardware. The mission managers for these seven programs would have preferred to start from a "clean piece of paper" but this would have made their particular mission too costly for the NASA budget. And when the hard decision came to have a mission using the MMS or no mission at all, one could predict the outcome: "We'll use MMS."

Well after the demise of the LCSO, the discipline of the budget still forced programmatic decisions along the standard spacecraft lines. It was not until the NASA budget growth of the late 80's and early 90's that the standard concept was abandoned. Arguments such as "technological obsolescence" and "old designs" then carried the day for a return to the "clean sheet of paper" concept.

There are still six NASA MMS based missions in orbit today. No MMS mission has failed. The cost effectiveness of the MMS concept is astonishing. Two Landsat spacecraft are in orbit today, producing data for 13 years and 11 years, respectively. The total cost per system was about $500 million dollars or so far about $100,000 per day.

Two major aspects of modularity were key to cost control of the MMS missions. The first was the flexible mission configuration. Second was rapid mission integration. Some 98 components would arrive at the mission contractor's plant in four separate modules, not one with the spacecraft structural frame; all were fully tested and ready to fly on any of a series of launch vehicles including Delta, Atlas, Titan and Shuttle. If, during testing, a component failed inside a module the entire module was replaced. And within 40 minutes the test was continued while the malfunctioning subsystem component was being repaired. The capability was critical to cost control for studies conducted in the early 70's at GSFC showed that fully one third of the entire mission costs were incurred during mission integration and test. By making access to the spacecraft quick and easy on the ground, the ability to repair it on orbit through modular replacement came virtually free. There is a small weight penalty for ground and orbit subsystem change-out capability, but it's a very favorable cost tradeoff for most systems.

Modularity provided the capability to upgrade technology from mission to mission and to replace on orbit an old technology subsystem with a new one. Such was the case with the Solar Max repair mission, where a control system was replaced with one of improved accuracy. When the replaced subsystem was returned to Earth, it was refurbished with new technology and flown on the UARS program, launched in 1992. The refurbished unit has been working flawlessly ever since. The first NASA "secondhand" spacecraft subsystem was refurbished at a cost of less than 25% of a new subsystem. The MMS modular concept of the 70's is in use in most of today's large spacecraft.

Frank Cepollina is still actively attempting to reduce costs through on-orbit servicing and maintenance. "Don't throw away your assets when they malfunction," he advises and based on NASA's past and present experiences in spacecraft servicing he makes a powerful economic case.

Cepollina feels strongly that the experience to date clearly shows that:

• Reuse or repair of space assets is smart business.

• NASA has demonstrated the ability to retrieve, refurbish and improve spacecraft as well as upgrading spacecraft on orbit.

The principles laid down a quarter century ago are still valid in today's NASA.

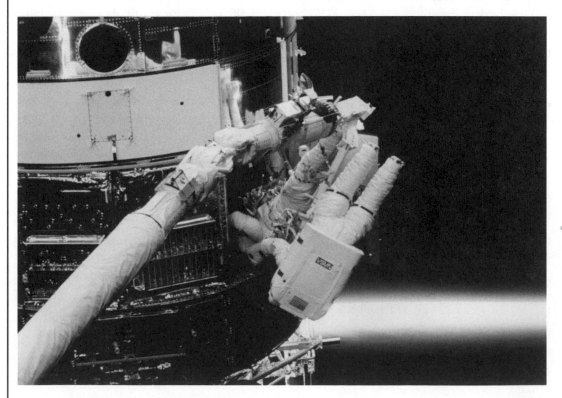

Hubble Space Telescope Repair Mission.   *Photo courtesy of NASA*

74

Footnotes

[1] A. O. Tischler, "Low Cost Space," paper presented to AIAA, ASME, SAE meetings. Denver, Colorado, July 10, 1973.

[2] George M. Low, Management Instruction: Standard Space Systems Program. Washington, D.C., NASA, October 9, 1974.

[3] A. O. Tischler, note to Frank Hoban, November 11, 1994.

[4] George Muinch, "Panel Activities," memorandum to NASA Centers, August 8, 1974.

[5] Edward Z. Gray, "Standard Equipment Procedures." Washington, D.C., NASA, May 5, 1975.

[6] Planning Research Corporation, "Study and Analysis of Low Cost Practices for Space Systems." Huntsville, Alabama, PRC D-2141, October 11, 1976.

Chapter 6

Changing the Way NASA Does Business

From the beginning of the low cost effort, it was obvious not just to Tischler but to others in NASA that significant savings could be found in changing the way NASA conducted its business. Hans Mark, the Director of the Ames Research Center, wrote to Ed Gray: "Studies have shown that hardware-oriented changes give savings ranging up to 50%, but factors of 10 or more can be saved through procedural (practices) changes."[1]

But change would not come easy. One of the legacies of the Apollo Program working against such change was the "bag of gold" syndrome—that there was always more money available to solve problems or to make things better or to buy insurance through tests and analysis. Aaron Cohen, the Orbiter Project Program Manager during the Space Shuttle's development phase, said that one of his most difficult problems was convincing his staff of Apollo veterans that there was no similar "bag of gold" for the Shuttle Program.

Left to right: President John F. Kennedy; Robert Semans; Dr. Wernher von Braun; James Webb, NASA; Vice President Lyndon B. Johnson; Secretary of Defense Robert S. McNamara; Dr. Jerome B. Wiesner, Science Advisor to the President; and Dr. Harold Brown.

Photo courtesy of NASA

This belief in endless supplies of money may have had its origins in the development of the estimate for the Apollo Program, when the numbers continually changed from $20 to $40 billion and back. James Webb, the NASA Administrator, reported to a Congressional Committee that "some people use a number as high as $40 billion to land a man on the moon. Others say half that amount,"[2] and way back in 1960 NASA had come up with an estimate of $26 to $38 billion. With a variation of that magnitude it was difficult to be too concerned about cost.

Industry was quick to understand that cost would not be a major factor in determining Apollo's success. In 1994, Joseph Gavin, Grumman Aerospace Corporation's Lunar Module Program Director and later Grumman's President and Chief Operating Officer, described the management of cost on Apollo:

> On paper, Grumman's contract with NASA allowed for some tradeoff between performance, schedule and cost. But it took us only a couple of months to learn that there really wasn't any tradeoff. You absolutely had to give priority to performance. Then you did the best you could to meet the schedule. Cost sort of came third. That may sound irresponsible, but when you think about it that's the way things had to be for Apollo.[3]

There were no incentives or even opportunity to do business in a more cost effective manner.

But there was hope. By the early 1970's, other government organizations, especially the Department of Defense, were striving to control cost, and many of their attempts were practices rather than hardware solutions. As John S. Foster, Director of Defense Research and Engineering, wrote to George Low on May 5, 1972:

> Increasing cost of our weapons systems is a matter of serious concern to the Department of Defense. These systems costs have been tripling every decade, and we must either stop this trend or accept fewer and fewer numbers of more expensive weapons.
>
> We are taking positive steps within DoD to gain control of the escalating costs of weapons systems. We are placing great emphasis on obtaining good project managers, giving them a chance to run their programs and to remain in the position long enough to get the job done. We are also demanding realistic tradeoffs where we balance cost and performance and schedules for our programs. Before we move from one step to another in our acquisition process, we will make certain by test demonstration that the program has passed critical progress milestones and we will not commit to production until the program has passed the final hardware test and evaluation. We must start with the engineers and managers if we are to successfully cut costs.[4]

DoD did much more than just talk about cost control. They returned to proto-typing and flyoff competitions, which proved very successful, as witnessed by the outstanding performance and longevity of the A-10 and F-16 aircraft programs. They introduced the Space Test Program (STP), a low cost approach to flying pay-loads, and began an acquisition cost evaluation program known as Project ACE. When NASA entered its low cost era, many aerospace contractors who worked for both NASA and DoD were already aware of a new business direction, and some had participated in serious cost reduction activities.

The Study Effort

In an effort to better understand the complexities of cost generation, mitigation and control, and thereby gather hard data on which to base recommended changes, the LCSO conducted extensive studies that focused on management, not hardware. The office sponsored, participated in, or otherwise was involved in over 30 studies related to cost identification, reduction or control. A number of these studies revealed significant findings and recommendations, and taken as a whole, con-tributed significantly to the expanding field of cost management.

The studies primarily concentrated on programmatic and technical cost dri-vers. Programmatic cost drivers focused on management and business practices that impacted cost and schedule; these ranged from documentation and contract management to risk analysis and safety. Technical studies covered specific engi-neering practices, including testing, systems engineering, design, production and standardization. Many believed these were the drivers responsible for the high cost of complex aerospace programs. The studies were conducted by both contractor and NASA personnel using diverse methodologies that contributed to the validity of the findings.

Practices Studied by the LCSO

Programmatic	**Technical**
Documentation	Specifications
Hardware Compendium	Testing
WBS Standardization	Spares Practices
Risk Analysis/Safety	Reliability and Quality Assurance
Program Management	Systems Engineering
Program Definition	Payloads
Contract Practices	Parts/Materials
Scheduling	Design Practices
Program Planning	Logistics
Organization/Management	Operations
Cost Control	Standard Components

A sample of the study results follows. Unfortunately, many of the findings and recommendations were never tested by NASA.

Boeing Space Test Program/ Mariner Venus Mercury (MVM73) Study

The "Feasibility Study of the Boeing Small Research Module Concept" was conducted to analyze the low cost techniques and practices used on two Air Force Space Test Program flights and compare them with those used on NASA's MVM73 project. All three projects had been judged highly successful, were accomplished in roughly the same time period and used the same contractor, Boeing, which provided a unique opportunity for comparison. The study was not a critique, nor did it attempt to define a best way; instead, it analyzed the data with the intent of explaining program differences.

The primary differences were: MVM73 had a larger customer organization (that is, more people in the project office), used a more detailed contractual task and subsystem definition process, and had a more extensive test program than did the Air Force STP programs.

The larger customer organization and work unit approach used by the Jet Propulsion Laboratory in managing MVM73 ensured the transfer of previous Mariner spacecraft experience from JPL to Boeing right down to the workers. This was thought necessary because much residual hardware from Mariner's '69 and '71 spacecraft was to be used on MVM73. JPL wanted to ensure that the lessons learned on previous Mariner missions would be applied on the MVM73 project. The disadvantage of this approach was the large number of one-to-one personal interfaces. Boeing was forced to greatly expand the size of its JPL project organization, this in contrast to the small Air Force STP team that required little interface management. The larger JPL organization affected the number and size of meetings, briefings, reports, planning activities and design reviews. Also program control, scheduling and bookkeeping activities were expanded to monitor and track individual work packages. Specifications and subsystem functional requirements were more detailed than for other typical STP programs due to the level of task and system breakdown necessary to incorporate the residual Mariner hardware and design data. The extent and level of testing on MVM73 far exceeded the typical STP program. Still there were a number of failures on this otherwise highly successful program.

In its final report, Boeing concluded that the following practices should reduce program costs:

- Use a small customer organization, oriented toward overall mission and design requirements, to permit the contractor to develop components and subsystems with minimum customer influence and reporting.

- Use top-level specifications of mission and/or spacecraft requirements to which the contractor must work with little further breakdown.

- Use an efficient but reasonable test program emphasizing system-level performance of flight hardware at expected flight environmental levels.[5]

The Aerospace Corporation Studies

The Aerospace Corporation conducted a study of the Air Force's experience in processing low cost spacecraft as part of the Space Test Program. The study cautioned that with the cost of payloads steadily increasing, more emphasis was being placed on achieving higher reliability and successful on-orbit performance. And in order to reduce risk, many customary practices might change along with a corresponding increase in cost.

The contractor found the most significant factors in reducing costs involved:

- Thoroughly screening payload requirements

- Procuring flight articles only

- Using "off-the-shelf" components

- Minimizing documentation

- Planning for tight schedules

- Using quality parts

- Using independent monitoring[6]

Another study completed on November 30, 1977, focused on practices that would improve program success and reduce the cost of spacecraft programs. No hardware-related practices were examined, only those associated with program management, systems engineering, quality assurance and testing. Twelve spacecraft programs (six NASA and six DoD) and 12 Space Test Program payloads formed the database. The 12 spacecraft programs represented nine contractors and were launched between the late 1960's and early 1970's.

To formulate cost reduction and successful improvement practices, the following activities and spacecraft characteristics were compared:

- Complexity—A definite correlation was found between increased spacecraft complexity and increased costs.

- Weight—No correlation was found between spacecraft weight and costs.

- Program Duration—Obviously, the longer the program duration, the higher the costs. Ways to reduce program duration were: designing spacecraft with maximum use of developed components, eliminating separate qualification spacecraft; and defining advanced technology items before the contract is initiated to minimize schedule stretch-outs.

- Program Management—It was found that the percentage of cost devoted to program management is directly related to program success. However, the correlation was only for the development phase, not for the production phase.

- Systems Engineering—This is the function of integrating subsystems, spacecraft, payload, and launch vehicle, and all technical activities not associated with subsystem design. Again, a direct correlation was found between technical success and systems engineering during the development phase, not during the production phase.

- Developed Hardware—Development costs were found to decrease by 30% as the percentage of developed components increased from zero to 100%.

In conclusion, the following practices were found to improve program success:

- Fully implementing program management and systems engineering during the development phase and maintaining a high level of quality assurance during the production phase. Performing a thorough acceptance test program and designing toward low complexity.

Practices that reduce cost without affecting program success included the following:

- Specifying spacecraft requirements that enable low-complexity design.

- Emphasizing reducing complexity and program duration rather than weight.

- Reducing management and system engineering activity during the production phase.

- Controlling quality assurance activity during the development phase by using contractor practices.

- Using the protoflight test concept and discontinue the prototype test concept.[7]

NASA Petrone Study

In early 1975, Rocco Petrone, the NASA Associate Administrator and Agency general manager for programs, asked the NASA community to provide him with new thoughts on reducing costs. The response was overwhelming. Some of the more interesting comments include the following, by category:

Requirements

- Initial Requirements are too ambitious. This is by far the most important reason for high costs. We are now traveling down the same road in the payloads business by busily inventing requirements that don't exist. The only answer to this is to actively push programs that do not have high costs built into them in the first place, [like] . . . airplanes, balloons, sounding rockets and inexpensive spacecraft programs.

- Maintain close working arrangements with the scientists involved in a project to search for ways to maximize science return within cost constraints or goals.

- Perhaps the greatest potential for low cost programs is in a complete and comprehensive definition of requirements prior to initiation of a program. NASA should require that each potential project demonstrate the achievement of a mature Phase B before proceeding into Phase C. Before commitment to Phase C, one should have established clear mission requirements, a preliminary design, the status of needed technology, a reasonable schedule, and a creditable cost estimate. Inflated requirements are perhaps the greatest drivers of costs.

Management

- Changes of direction during the program. This has been cited frequently as being an important cause of high costs. . . . The essential point here is that once the project is defined, Headquarters and the Center must back the Project Manager in resisting changes. Often, the Project Manager does not have sufficient political clout to overcome these pressures without outside help.

- Mismanagement. We all know what this means but the two most common failings are bad technical judgments and mismanaging contractors by not motivating them properly. The only remedy for the former is not to keep people who have proven themselves incompetent.

- We tend in general to over-manage our contractors. . . . NASA should cut back the number of management levels and degree of checks and balances. The high level of inhouse expertise necessary for Apollo is just too luxurious for our future.

- Business management reporting and data collection systems are too cumbersome and duplicative.

- Require program planning to mature prior to major contractor manpower buildup or initiation of procurement actions. Contract implementation should provide for a sufficient period of cost tradeoffs before initiating the in-depth design activities.

Standardization

- Promote standardization and commonality techniques and the broad use of proven flight hardware and create an atmosphere conducive to the development of common systems and hardware that is close to the mission objective (simple in design and built only to the complexity level required).

- Standardize at the highest practical level to achieve maximum cost savings.

- An important factor in cutting costs is the ability to make quantity buys. This reaps the advantage of lower prices because of quantity production, but if applied properly, can greatly reduce lead times.

- Scrounge for flight-qualified parts and equipment throughout the government.

- Share intercenter and interagency facilities.

Sensor and Instrument Development

- Sensor and instrument development should be done by those persons or organizations who conceptualize the idea, and they should remain fully involved and committed to that device through its flight operations.

- Technological advances between flights should be maintained at modest levels. If a new sensor or instrument is needed, it should be developed as a part of supporting research and technology, not project activities.

Administrative Processes

- Continue reducing reporting requirements at all levels.

- Obtain any equipment which, in the long run, automates systems, thereby reducing or eliminating people.

- Establish meaningful rewards for real cost-saving ideas.

- Consider initiation of formal cost reviews with contractors concurrent with conceptual and initial design reviews. The cost review can be conducted on the same NASA/contractor management level as Design Acceptance Reviews.

Other Cost Saving Ideas

- Instill in every employee the firm conviction that costs must be reduced at every level.

- Conduct a critical review of Agencywide purchasing and contracting procedures and policies with cost-saving the goal. Consider foregoing the use of government specifications for items not critical in a safety sense.

- Be prepared to end the project if costs get out of hand. A real motivator.

- In addition to using existing designs where possible, there is no substitute for experience. The maintenance of inhouse competence is essential.

- Consider creating a full-time technical organization at each major NASA installation and within each major program whose primary duty is to find areas to save money.[8]

These and dozens of other worthwhile suggestions were collected, analyzed, grouped and circulated to NASA management. But Petrone and Low clashed on other issues. Petrone left NASA and the study recommendations were never implemented.

Systan's Study of Cost Overruns

A particularly interesting study with the unwieldy title of "Contract Performance on Unmanned Spacecraft Projects: Effect of Prime-Subcontractor Relationship" was conducted in 1975 by the Systan Corporation. The study examined such relationships in Ames Research Center's and Goddard Space Flight Center's unmanned projects. Systan Corp. noted that "the major problem in assessing overall contract performance is in measuring cost performance." In the case of a cost overrun, they argued the necessity of analyzing both the official cost overrun (based on the contractor's cost estimate and subsequent contract award) and an unofficial overrun (based on NASA's estimate of the expected cost) to properly analyze the cost performance. Both figures are necessary since some contracts with official cost overruns and unrealistically low bid prices may or may not be unofficial cost overruns. "That is, contract cost performance may be good when

Nov. 3, 1993

Robert A. Frosch, testimony to the Senate Committee on Commerce Science and Transportation Subcommittee on Science Technology and Space —Washington, DC

"It is sensible to carry out assignments as efficiently and frugally as makes sense given the nature of the assignment, but it is not sensible to agree to, or to be forced to agree, to do a project much more cheaply than anyone really believes is possible."

85

Cost Control Measures Aimed at Saving Money on International Station

Washington – NASA is introducing a number of measures to control costs on the international space station, ranging from using fixed-price contracts to reining in money that is still being spent on a previous design.

Space News—April 11-17, 1994, p. 25.

allowance is made for the buy-in." It is interesting to note that as late as 1992, high-level NASA officials were still denying the existence of buy-ins. But most NASA project personnel were aware that buy-ins occurred at all levels of NASA management and sometimes even at the Congressional level, in addition to what the contractors were guilty of. The Systan study remarked: "Even with both sets of overrun figures, analysis of contract cost performance often would not be very meaningful because neither industry nor NASA has developed great accuracy in forecasting contract costs, particularly the more complex and technically uncertain projects." Systan stated that Ames and Goddard had increased the use of fixed price contracts in an effort to limit NASA's cost liability, noting that

> fixed priced contracting is particularly appropriate when processing follow-on spacecraft in a series; it is more appropriate for procuring application satellites with well-defined missions and well-known technology than for procuring scientific or research satellites with less well-defined terms of reference and for which a greater proportion of technical development is required.[9]

The rationale is that for fixed priced contracts, the contractor bears all the risks of cost overruns, and therefore will estimate costs more closely and submit a bid with a large contingency. As such, the fixed price contract may have a higher initial price than similar cost reimbursable contracts. Fixed price contracts can be increased for scope and engineering changes and they discourage buy-ins but they do not prevent them. Theoretically, fixed price contracts would allow NASA to know its expected program costs with more certainty.

In order to discourage unrealistically low bids, the study suggested that NASA specify an estimated range of effort that the successful completion of the contract would require. This estimate, developed inhouse, should itself be an honest attempt to understand total cost and should not contribute to the buy-in phenomenon.

NASA, Systan asserted, can characterize contractors based on their past performance. Even though these characterizations can fluctuate over time, there was a general agreement as to low cost producers, the strong versus the weak, the responsive and unresponsive firms, and an association with a particular type of spacecraft. However, NASA does not contract exclusively with the "best" contractors, primarily because of the need to maintain competition and the fear of being left with only a few firms capable of building spacecraft.

A final Systan recommendation was that NASA should decide what a particular program is worth in terms of cost and benefits, and manage to that value. An upper limit of program costs and an assessment of a contract's estimated cost at completion should provide the necessary backbone to keep costs down and also provide the basis for project cancellation if costs should exceed the established value. The earlier this is done in the program, the more credible the threat.

Lockheed's Study of NASA Practices

In May 1974 Lockheed Missile and Space Company completed a four-year study of the design, manufacturing and on-orbit maintenance techniques used in creating low-cost payloads for the Space Shuttle, as well as the design of low-cost space hardware in the areas of standardization, modularity and low-cost manufacturing techniques. The underlying philosophy in this report was twofold:

1. If a program requirement is desirable but not mandatory, eliminate it.

2. If a program requirement is mandatory, reduce it to the minimum level consistent with the required capability.

Lockheed's stated purpose was to identify and quantify those routine NASA practices that cause the costs of space programs to increase. Top cost drivers were found to be:

1. Too many people at the interface, with a high ratio of government-to-contractor personnel

2. Insufficient early project definition

3. Delays in selecting experiments/payloads

4. Divergent cost-estimating methodologies

5. DoD procurement processes that did not match NASA's business practices

6. Too many high-reliability/quality control requirements

7. Excess documentation and extended approval cycles

8. Overly stringent specification requirements at lower procurement tiers, and

9. Changes and schedule slippages.

Cost comparisons indicated there could be a typical savings of 27% if low cost techniques replaced "routine" ones on future projects. The four areas in which low cost recommendations were made were: program specifications, program documentation, general contracting, and program definition.

Pentagon Revises Arms Acquisition Rules

The order went out in a May 10 letter from Defense Secretary William J. Perry, that program managers could give some preference in future contracts to companies that have proved efficient and trustworthy in the past.

"Perhaps the surprising thing is that we have not been doing this in the past," Kaminski said of the Pentagon's cumbersome oversight process that too often has discovered problems in programs long after they should have been found.

The Washington Post—May 21, 1995.

1. Program Specifications

Two specifications were selected for detailed study as to their effect on cost: 1) NHB 5300.1, which establishes NASA's reliability and quality assurance standards, and 2) MIL-D-1000, Engineering Drawing Practices. From these two studies, Lockheed concluded:

The NASA handbook series should be replaced with mandatory minimum requirements.

Considerable cost savings are possible by using only one level of MIL-D-1000 instead of several.

"Hazard Free" hardware can cost 35 to 40% more than hardware without that rating and should be required only where essential.

NASA and its Field Centers need a strong specification control office and a cost-effectiveness evaluation should be performed for all future specifications generated. Specification language and cross referencing should be simplified and a standard specification format is needed.

2. Program Documentation

Documentation accounts for 20% of total program cost; it could be cut in half by altering present practices to a mandatory-minimum level. For instance, drawing changes cost twice the original drawing costs; these are usually attributed to a lack of early program definition and contractor errors.

3. General Contracting

Approximately two-thirds of every project dollar is spent on labor and labor benefits; more management attention must be focused on limiting and/or eliminating people tasks.

Reduce contract regulations, directives, specifications, and requirements to mandatory minimums by reworking or replacing existing ones.

Change the dollar range of incentives for Cost Plus Incentive Fee (CPIF) and Cost Plus Award Fee (CPAF) contracts or eliminate these contracts entirely. They cost much more to administer than other contracts and divert attention from more important cost-impact aspects of a project.

Flexibility to make late changes/improvements in hardware must be traded for a low-cost approach that requires early and complete program definition based on known technology, maximum use of standardization and modularization of hardware, and a mandatory-minimum management philosophy.

4. Program Definition

Separate new technology development from the overall spacecraft hardware. This will reduce changes and cost growth during Phase C; it will reduce risks and delays; and it will improve program definition for hardware elements.

Emphasize early and complete program definition.

Base estimates on realistic costs, but have a reserve fund for uncertain areas.[10]

Lockheed's four-year study effort made significant contributions to NASA's understanding of costs.

OAST's "Experimental Shop" Approach

A study initiated by the NASA Office of Aeronautics and Space Technology (OAST) led to a *Guide to Management of OAST Experimental Programs* ("Experimental Shop" Concept), dated June 4, 1973. This guide outlined management practices to reduce the cost of one-of-a-kind programs using an "experimental shop" approach, in which costs were given equal consideration to the technical aspects of a project. This approach is characterized by:

- A small, highly qualified government and contractor project organization with a close relationship among engineering, fabrication, and test;

- A good work breakdown structure (WBS) keyed to planned costs;

- A mutual understanding between the contractor and OAST of the project requirements, achieved in part by having a "60-day introductory period" after contract start to assure complete understanding;

- Close, continuing communications between the government and contractor, including visibility into the major subcontracts as to cost, schedule, and technical progress;

- Reduced documentation, reporting, and inspections;

- Simplified change controls; and

- Mutual motivation to control costs.[11]

This experimental shop was used on small development projects, but was never used on a large planetary or crewed program.

Air Force ACE Project

Project ACE (Acquisition Cost Evaluation) was an effort on the part of Air Force Systems Command to reduce acquisition costs. The Air Force was very concerned about the rapidly rising cost of weapon systems coupled with the decrease in their share of the Department of Defense budget. In 1972 Systems Command initiated an evaluation to determine opportunities for cost reduction. Workshops were held in the spring of 1973 that further defined the opportunities, and an ongoing program was launched. It was in full swing by the time NASA launched the LCSO. Some of the topics of common interest were:

- Attention to development programs in the pre-acquisition phase

- Use of tradeoff analyses

- Development of tools and techniques for life cycle costing (LCC)

- Development of a corporate memory

- Procurement planning and acquisition strategy

- Quality assurance

- Development of experienced program managers

- Design-to-cost implementation[12]

Project ACE should have had more impact on NASA's low cost effort. Unfortunately, in the initial Air Force briefing to NASA, the project manager, a young full colonel, made a fatal mistake. Low previously had a very detailed briefing on the A-10 close support aircraft from the Fairchild Aircraft work force. He knew a lot about the program. The A-10 was on the ACE briefing agenda. After just a few questions it was obvious to the attendees that Low knew much more about the program than the Air Force colonel did whose responses were for the most part inaccurate and uninformed. Based on this lack of knowledge, Low discounted the entire effort.

NASA's Inhouse Study of Business Practices

In early 1975, the Business Practices staff concluded an inhouse study with Headquarters offices and Field Centers. The respondents agreed on two issues, each close to the results of the contractor study findings:

- NASA does not know what things should cost or why they cost what they ultimately pay.

- NASA's major cost problems arise from poor definition and unrealistic requirements and more resources, both contractor and inhouse, need to be spent in the definition phase of the project cycle.[13]

The study also found that budget reductions in supporting research and development had forced projects into concurrent technology development with resultant higher costs and project schedule slippage. A return to some form of a technology test flight program was recommended.

As for risk taking to reduce cost, the Delta Launch Vehicle project experience continued to haunt NASA. The Delta team took pride in operating in what they thought was a "low cost mode." After years of phenomenal success, the team eliminated or reduced testing, documentation and launch instrumentation in a determined attempt to keep Delta "low cost" and competitive. The project then suffered two failures and NASA management responded with a much feared Failure Review Board. Many at GSFC felt that the Board's report was extremely critical of the very steps the project managers had taken to keep Delta a low cost vehicle. The project manager was relieved of duty and eventually left NASA. The fear of "a similar Review Board if you fail" was evident at all NASA Centers. Most project managers interviewed were unwilling to take risks to reduce costs because they feared management would not stand behind them if a failure should occur. Project managers admitted to "buying insurance" in the form of additional analysis and tests to protect themselves from failure reviews, knowing full well the upward impact on cost.

And lastly, NASA had prided itself on being a "can do," dynamic, success-oriented agency, and the reward system followed this theme. Cost control or reduction had never been a criterion for success. NASA managers were raised on the maxim, "They'll forgive you if you overrun but they'll never forgive you if you fail." They were not inclined to worry about cost.

Goldin Subjects Cassini to Cost, Risk Reductions

"We can't fail with that mission. It would be very, very damaging for the agency," Cordova said March 9.

Space News— March 14, 1994.

Low Cost System Office Studies

Title of Study	Researcher
• Petrone Study	NASA personnel
• Program Practices Analysis (Boeing Small Research Module)	Boeing Aerospace
• Project ACE: Findings and Actions	Air Force Systems Command
• Cost Analysis Study	Ball Aerospace
• Experience in Procuring Low Cost Spacecraft	Aerospace Corporation
• Low Cost Program Practice	Lockheed Missile and Space
• Mission Operations Low Cost Study	Jet Propulsion Laboratory
• Atmospheric Explorer Low Cost Study Report	RCA
• Standardization and Program Practice Analysis	Aerospace Corporation
• Contract Performance on Unmanned Spacecraft Projects: Effect of Prime-Subcontractor Relationship	Systan, Inc.
• Report of the Task Force on Reducing Costs of Defense Systems Acquisition, Design to Cost Commercial Practice vs. Department of Defense Practice	Defense Science Board
• Report to the NASA Space Cost Evaluation Program	NASA Historical Costs Panel
• Comparative Analysis of HCMM/SAG Program Practice	Boeing Aerospace
• Program Practices Analysis	Aerospace Corporation
• Data Collection for Unmanned Satellites	General Electric
• Cost Study of OSO-1	Hughes Aircraft
• ATS F Cost Study	Fairchild Space and Electronics
• Cost Analysis of the ITOS, D,E,F, G & E2 Spacecraft	RCA
• Study to Determine the Effectiveness of Temperature Cycling vs. Thermal Vacuum Testing of Spacecraft	International Mgmt. Resources
• Cost Benefit Assessment of Standard Equipment, Program Practices and Business Practices	NASA
• Environmental Design Requirements and Test Criteria for Standard Components	NASA
• Environmental Test Methods and Controls for Components	NASA
• NASA Composite Index of Specifications	NASA
• Study and Analysis of Low Cost Practices for Space Systems	Planning Research Corporation
• Equipment Specification Cost Effect Study	RCA
• Low Cost Program Practices for Future NASA Space Programs	Lockheed Missile and Space Corporation
• Mariner, Venus, Mercury 1973, A Study of Cost Control	NASA
• Report on Cost Estimation and Control Seminar	NASA
• Cost Benefit Analysis of Standard Equipment	NASA
• Advanced Systems Cost Estimating Techniques	Planning Research Corporation
• Standardization and Program Effect Analysis	Aerospace Corporation
• Pressure Vessel Spacecraft (PVS), Multi-mission Application	General Electric

Other Activities of the Business Practices Staff

In addition to these studies, a host of other activities were conducted by the practices group. Shortly after the office was formed, the staff looked for help in getting started; they wanted to know where and how NASA should change the way it did business. Most respondents were open and helpful. Others did not disguise their belief that cost reduction was merely "smoke and mirrors" to impress the Administration and the Congress. But there was some support for change, including change not directly associated with cost reduction. Many believed that post-Apollo NASA would be different, but they did not believe the change would be as easy as Low suggested in his speeches.

One of the first obstacles they ran into was the question of standard equipment. This was appropriate, in that a move toward standardization was a major change in the way NASA did business. The issues raised were simple and to the point. Why should Field Centers be involved in standard equipment? Who would be blamed if the standard equipment failed? How would the LCSO effort not interfere with the basic mission of the project? These questions were asked in the first months of this effort. Years later the LCSO was still working on acceptable answers.

The practices staff conducted an ambitious program that lasted more than six years. Its accomplishments were prodigious: project management workshops (Chapter 7), the first Agencywide project management training program (Chapter 9), the CASH catalogue (Chapter 5), the first understanding of Agency culture (Chapter 10), and the development and promulgation of a payload policy and the project management review process (Chapter 11).

In addition to these accomplishments, many other studies, analyses and initiatives were undertaken. Chief among these were the NASA acquisition and financial management processes, the DoD Space Test Program, NASA specifications and drawing requirements, project management practices and cosponsoring the introduction to NASA of quality circles.

Initiatives in support of the NASA financial management function included a standard financial system, a standard centralized property management system, and a payroll and personnel system by which Headquarters and Centers would share and pool resources in developing common operating systems.

One of the first and most successful of their undertakings was the conduct of a reports "murder-board." This initiative was in response to numerous complaints by Field Centers of excessive reporting requirements on the part of NASA Headquarters. Approximately 150 reports were eliminated with annually recurring savings of $1.5 million. The entire process was so efficient and effective as to be a model for future activities.

The LCSO looked to overhaul NASA's acquisition processes. One of the most successful efforts was the computer-aided procurement management technology program, which was designed to create Agencywide procurement systems. These systems would improve the efficiency and effectiveness of procurement operations by using uniform computer systems and minimizing interface problems between customers and the procurement organizations. There were five modules in the pro-

Hunting Sacred Cows

In a paper "Lessons Learned from Hunting Sacred Cows" delivered at the Fourth International Symposium on Space Mission Operations and Ground Data Systems, September 16-20, 1996, in Munich, Germany, Rhoda Shaller Hornstein reported on the insights and discoveries her NASA cost less team made during their recent deliberations. The first of these was that NASA must change the way it does business from a smart builder of one of a kind systems to a smart open market buyer.

The More Things Change

The headline in a Dec. 6, 1993, issue of Federal Computer Week could have been written two decades earlier: "NASA Proposes Guidelines for Acquisition Streamlining." NASA's office of procurement policy had proposed a set of techniques to shorten and simplify the contracting process while cutting costs.

gram: purchase request tracking, cost modeling, cost and pricing data bank, procurement instrument generation and bidder's list.

In 1975 an extensive procurement review was undertaken. Items reviewed included indirect costs, award fees, service contracting, procurement lead times, RFP process, source selection, buy-ins and noncompetitive justification. Although much time and effort were spent reviewing these topics, only minor changes were made: procurement management support was less than wholehearted. It was later learned that the procurement leadership instructed their staffs to game it. At this same time, a major study of the Source Evaluation Board Process was conducted. Two recommendations were noteworthy; the number of SEB members should be kept to a minimum and each Center should establish a more formal method of indoctrinating (training) new members.

The DoD Space Test Program (STP) was of great interest to NASA. It was well known as a low cost program even before Low initiated the NASA effort. The basic tenets included: a small project office of usually no more than five people, a condensed test program, minimum customer influence and reporting, and little breakdown of tasks and work packages. A good description of the program was written by Del Tischler on June 8, 1973.

> STP is a low cost space program which bears some distinct similarities to the efforts undertaken in NASA to cut space development and operational costs. It comprises a series of space launches involving direct or piggyback spacecraft and experiments. STP does not cover the experiments or scientific payloads. These are supplied and paid for by user agencies.
>
> Launch vehicles are often "launch vehicles of opportunity." Outdated Atlas F vehicles are being rejuvenated for dedicated flights. STP is flying about two or three flights per year although there are plans for some 70 future payloads. The launch success ratio on Atlas F is 91% (which is above reliability predictions).
>
> The spacecraft (not including experiments) cost on the order of $3-10 million, averaging $5 million. Equipment standardization has not been foremost in getting this relatively low cost. Instead, costs are given prime emphasis, which in turn has led to simple design and effective use of proven equipment.
>
> What impressed me was the enthusiasm of the payload-experiment (user) groups for the launch and spacecraft services. Here is an instance of clear separation of the payload user from the organization responsible for both the spacecraft and launch vehicle, and it works in spite of the integration problem in the spacecraft/payload area. The solution seems to be predefining the experiment interface characteristics before soliciting bids for the spacecraft. The spacecraft contractor becomes responsible for this integration. Also, a general interface document is supplied to the user.

STP has guidelines which parallel our low cost system effort:

- Utilizing previously flight-proven/flight-qualified hardware.

- Utilizing low-cost launch vehicle systems.

- Rigorously negotiating payload "desirements" until well defined "requirements" are established.

- Procuring competitively (if appropriate).

Total development and operation schedules are short, covering 24-27 months total time from announcement of flight opportunity to flight.

The STP staff is 22 people. They draw on The Aerospace Corporation technical or costing experts, which number as many as 50, but do so only upon necessity.

The group seems intent on improving their cost estimating—almost always done within this small group—in order to contract on a design-to-price basis. The experimental test prototype is almost always the spacecraft actually flown.[14]

A study on Bill Stroud's Pressure Vessel Spacecraft was completed to evaluate cost savings to be had by using spacecraft housekeeping equipment housed and operated within a "tank" pressurized with an inert gas. The spacecraft size would grow considerably but the cost savings were estimated to be significant; the pressurized environment would permit the use of commercial equipment without significant modification.

The Drawing Requirements Study found three forms of deliverable drawings that varied significantly in total cost. Form 1 drawings were to military standards. Form 2 drawings were to industry standards with partial military controls and Form 3 drawings were to industry standards with minimum military controls. Form 1 cost twice as much as Form 3. Of 62 NASA contracts, all over $2.5 million in total value, approximately one-third were Form 3, one-third were Form 2, three were Form 1 and the rest were unspecified or did not require drawings. The recommendation was for NASA to specify Form 3 in future procurements.

A specification review revealed that to reduce costs in future programs, NASA should reduce the number of applicable specifications, simplify specifications and make program managers aware of the cost implications of imposing a particular specification so that cost/risk tradeoffs could be effectively carried out. A NASA composite specification index was prepared and several specifications were rewritten as low cost examples.

Overall, the practices effort was very successful. The extent to which it permeated the NASA culture is difficult to evaluate, since there is no way to know how many memos similar to the following flowed through the work force. The following excerpts are from a memo to Fred Speer, the HEAO-C project manager from Horton Webb, a member of his staff:

> I feel we can successfully apply some of the low cost program practices to HEAO, particularly to the HEAO-C experiment contracts which are now being developed.
>
> . . . We should organize a small ad hoc group that is strongly motivated to be cost effective. The group should consider but not limit itself to:
>
> • Changing drawings from Form 2 (Government Standards) to Form 3 (Contractor's Standards).
>
> • Deleting some of the deliverable documentation.
>
> • Greatly reducing the number of government specifications to be followed, and specifying precisely the application of the specs that are selected in order to avoid inadvertent application of second and third tier specs. Example of one that can be eliminated in experiment contracts: MIL-STD-130, Identification and Marking of U.S. Military Property, because there is no military property involved in the experiment effort.
>
> • Reducing the HEAO requirements where practicable, such as part of Management Requirements for Experiments—do we really need a Certificate of Flight Worthiness?
>
> For HEAO-A and -B contracts, I feel we should see how much of the HEAO-C reductions can be applied. The reductions will have to be sizeable to offset all the effort involved in revising the contracts, contract documents, and negotiating a cost reduction with the contractors.[15]

On April 1, 1974 Glynn Lunney, the Apollo/Soyuz spacecraft manager, wrote to his Headquarters counterpart in part:

> In February 1974, Rockwell International took delivery of the last VHF/FM transceivers.
>
> The approach used in procuring the VHF/FM transceiver was different from the normal Apollo program procurement. The objective of this "mini spec" procurement was to provide quality hardware at a minimum cost to the Government. This procurement was for seven flight units and one engineering model. Each unit consists of two complete transceivers.

The cost to the Government for these eight units was approximately $148,000. In addition to this, Rockwell performed a certification test. The initial cost estimate by Rockwell was $2.2 million. This estimate was based on the normal Apollo procurement and included the integration cost.

. . . We feel that this procurement has met the objective of providing quality flight hardware at a minimum cost.[16]

And according to Perry Westmoreland who managed the procurement and put the mini-spec together, the success was based on:

No NASA contract monitoring. Leave the contractor alone to do his job.

Select dedicated "low cost believers" as personnel for preparation of mini-specs.

Minimize changes on the program. This program had only one change of approximately $5000.

Throw away all "boilerplate" [standard contract language] material.

Write the procurement spec; in fact, write all mini-specs used to produce hardware or procure hardware on a by-case basis. Insist that the person responsible for the hardware/program element write the spec, and defend it against all comers.[17]

Westmoreland admitted that he wasn't overwhelmed by converts or believers, but no one could quarrel with the fact that the hardware worked successfully and did the job.

Although the practices effort continued to yield valuable cost savings ideas and methodologies, it never surpassed standard equipment in terms of resources or management attention. Tischler was right: engineers and scientists (and therefore, the vast majority of NASA management) simply liked to tinker with the hardware.

Footnotes

[1] Hans Mark, letter to Ed Gray, June 17, 1975.

[2] Howard E. McCurdy and John M. Low, "The Cost of Space Flight," unpublished article. School of Public Affairs, The American University, Washington, DC, May 1, 1994.

[3] "Fly Me to the Moon: An interview with Joseph G. Gavin, Jr.," Technology Review, July, 1994, pp. 61-68.

[4] John S. Foster, Jr., letter to George M. Low. Washington, D.C., Department of Defense, May 5, 1972.

[5] B. C. Padrick, "Feasibility Study of the Boeing Small Research Module (BSRM) Concept." Seattle: Boeing Aerospace, August 1975.

[6] Systems Engineering Operations, "Standardization and Program Practice Analysis." El Segundo, The Aerospace Corporation, December 15, 1976.

[7] "Program Practices Analysis." El Segundo, The Aerospace Corporation, October 6, 1977.

[8] Rocco A. Petrone, "Center Director Suggestions on Cost Savings." Washington, D.C., NASA, March 3, 1975.

[9] "Contract Performance on Unmanned Spacecraft Projects: Effect of Prime-Subcontractor Relationship." Los Altos, California, Systan, Inc., April 1975.

[10] "Low Cost Program Practices for Future NASA Space Programs." Sunnyvale, Lockheed Missile & Space Co., May 30, 1974.

[11] Office of Aeronautics and Space Technology, "A Guide to Management of OAST Experimental Programs." Washington, D.C., NASA, June 4, 1973.

[12] Air Force Systems Command, "Project ACE Findings and Progress Report." Andrews AFB, Maryland, U.S. Air Force, June 1974.

[13] "Proposed Business Practices Action Plan." Washington, D.C., NASA, March 5, 1975.

[14] Del Tischler, memorandum for the record, June 8, 1973.

[15] Horton Webb, memorandum to Fred Speer, November 12, 1974.

[16] Glynn Lunney, memorandum to Chester M. Lee, April 1, 1974.

[17] Perry Westmoreland, memorandum to Norman Rafel, November 14, 1974.

Chapter 7

Project Management Workshops:
Doers Share Experiences

Another key to significant change in NASA was the manner in which its work force managed projects. Low certainly understood this. He made a special attempt at the earliest days of the low cost effort to get their recommendations and advice; the "doers," as he described them, "will make happen what I have been preaching."

The practices group of the LCSO agreed with Low and wanted to engage the project management work force in an open discussion of their experience in identifying and controlling costs. This was assured on October 10, 1974, when Low met with the Aeronautics and Space Engineering Board (ASEB) of the National Research Council.[1] The Board, established in 1967 at the request of NASA, was to "provide conclusive reports on questions of national importance related to aerospace engineering—at the request of government agencies and on its own initiative." Low had asked the ASEB to establish an ad hoc study group to look at NASA's Mariner-Venus project as to cost drivers and potential areas for cost savings on future projects. The study was not going well and its participants were looking for direction.

After considerable discussion it was decided that the ASEB together with the LCSO would sponsor workshops to explore in depth with project managers the questions of cost generation and control. Two joint NASA/ASEB Project Management Workshops were conducted in Washington, D.C. in 1975; 15 NASA project managers participated in the sessions. Held at the National Academy of Sciences, the workshops identified cost drivers, but they also enabled the attendees to share various approaches employed in managing NASA's complex programs.

The ASEB was represented by retired NASA executives and Apollo pioneers Robert Gilruth and Abe Silverstein, retired USAF General William King, and Sid Metzger, Assistant Vice President and Chief Scientist of COMSAT.

First Workshop
The first workshop, held on February 24-25, 1975, covered nine projects especially selected to capture the full range of recent NASA management experiences.[2]

Gilruth opened the session on behalf of the ASEB, reminding the attendees they were to identify project cost drivers, discuss their experiences in handling the drivers, and using hindsight, identify what could have been done to further reduce costs. They were also to recommend cost control techniques for future projects.

The Projects

The Atmospheric Explorer Project consisted of three Earth orbital missions, each utilizing a spacecraft of approximately 1,500 pounds with payloads of approximately 200 pounds. The science objectives were to investigate the proton chemical process accompanying the absorption of solar ultraviolet radiation in Earth's atmosphere by making closely coordinating measurements of the reacting constituents from the spacecraft. The spacecraft was placed in orbit by the Delta launch vehicle. The project staff never exceeded 14 GSFC employees. The orbital mechanics of the mission permitted an unrestricted launch window, and the launch dates were met within 30 days of the target. David Grimes, the GSFC project manager, offered the following overall cost control techniques: spread project subsystems throughout the industry, thereby lessening overall risk; do not keep too many subsystems with the prime contractor; have the prime contractor use fixed price contracts where possible; motivate the contractor to keep costs low; and ensure that the project office and the contractor accept one leader, the project manager, for all elements of the project.

Grimes had specific recommendations for each element of the project team:

Recommendations:

For Contractors

1. Be willing to work as part of a NASA/contractor team rather than at arm's length and be extremely cost conscious.

For Project Managers

2. Get good people on the project team and make sure they talk to each other; instill in them the conviction that success can be achieved.

3. Be obsessed with cost and schedule—count things. Keep encouraging and pushing your people.

4. Maintain an information net that alerts you to difficulties within one day.

For Field Center Management

5. Ensure that the project leader has effective control of the inhouse manpower and project personnel and ensure a continuity of assignment of people to the project team.

The Mariner Venus/Mercury '73 Project consisted of a single spacecraft launch to the planets Venus and Mercury during the 1973 launch opportunity. The mission's primary objective specified a flyby of the planet Venus with a continuing trajectory toward a flyby of Mercury. Subsequent post-Mercury planning allowed for return encounters of the spacecraft with Mercury. The program had a firm not-to-exceed budget of $98 million with the stipulation that a spacecraft system contractor was to be used for the design, fabrication, and test of the flight spacecraft and test articles. The experiments and the participation of science teams were also limited to a fixed budget included in the $98 million ceiling. The project experienced excellent cost control throughout and underran the contract effort. The Jet Propulsion Laboratory's inhouse effort—consisting primarily of mission operations, tracking, data acquisition and science management—also experienced an appreciable underrun. Project Manager Gene Giberson elaborated on the guidelines used by his team during the management of the project:

- Establish firm inhouse mission specifications and strongly resist any deviation from them.

- Establish firm science mission requirements, including all science interfaces prior to spacecraft design.

- Establish firm cost estimates with principal investigators, and instill within the science team the not-to-exceed philosophy of the project.

- Establish a design carryover attitude for the subsystem managers and resist any state-of-the-art improvements.

A major point touched on during the discussion was the tradeoff between the spacecraft phasing alternatives available and the spacecraft systems contractor. One plan had the contractor work force building up rapidly, with the contractor buying all parts, completing all design effort and subsystem fabrication early before retrenching into a one year slack period prior to a second manpower buildup for final assembly, test and launch operations. This plan had the obvious advantage of staying ahead of the inflation spiral by completing all costly procurements early in the program. The second plan involved delaying contractor start as late as possible, building up fast, reaching a peak level of effort just prior to final checkout and launch, and then terminating the project activities in a very short period of time. The latter plan, adopted by the project, was considered success-oriented, but assumed considerable risk. It was recognized that this plan might not be the best approach for a program involving major new developments. Giberson reported these activities related to project success:

Pre-Project Mission Design

- Establish mission objectives and a science steering group.

- Regarding the science, mission and spacecraft design interaction:

  – Establish technical requirements/performance trades and develop preliminary cost estimates.

- Emphasize design carryover approach and conduct cost tradeoff analysis.

- Select "baseline" system configuration and establish target cost.

Project Definition and Planning

- Restrain staff size.

- Expand "baseline" system designs and interfaces and develop detailed cost estimates for implementation alternatives.

- Establish project guidelines and constraints.

- Conduct scheduling/cost trades:

  – Maximize cost predictability and control.

- Establish operating budget using a fixed-cost/variable-scope approach, emphasizing cost-at-completion and using no-year funds approach. Assure compatibility of scope and resources and stress candor on plans, allocations, and status.

- Prepare detailed implementation plans:

  – Prepare specific and detailed request for proposals and make careful make/buy tradeoff assessments. Use existing documents and administration systems and select a fee approach.

- Indoctrinate personnel:

  – Raise cost consciousness, make cost goal believable and foster an understanding of cost control plans and system.

Project Implementation

- Define contracts prior to start of work, maintain implementation and budget plans, do only essential work, and on-load and off-load manpower in timely fashion.

- Use Tiger Team problem solving approach (team of specialists to focus on issues of concern).

- Tailor test activities.

Recommendations:

1. Plan early and in detail.

2. "Start" late.

3. Use existing designs where practicable.

4. Establish cost-at-completion budgeting and control.

5. Communicate often.

6. Do only what's essential.

SPHINX was the smallest spacecraft discussed during the workshop. The objectives of the project were to obtain engineering data on the interaction between a high-voltage surface and space plasma. Although a launch vehicle failure terminated the operational phase of the satellite, SPHINX was considered successful from the standpoint of cost control and schedule performance. From its inception, the project was considered to be a high-risk, low cost effort (approximately $1 million), with no redundancy in the spacecraft.

An engineering model and a protoflight model spacecraft were designed, fabricated and tested inhouse. The experiment, a technically difficult, high-voltage instrument package, was designed and fabricated under contract.

Many problems were encountered during the contractual effort including difficulties in developing the high voltage instruments, lack of Center engineering support during the early part of the program, unavailability of parts, and the use of research contractor personnel for spacecraft support.

Recommendations:

1. Establish a realistic schedule early in the program.

2. Apply sufficient inhouse engineering design effort during the preliminary design phase.

3. Obtain a complete parts inventory as early as possible. If all parts are not available, make the design compatible with the parts that are obtainable.

4. Insist on project, not research, personnel from the contractor.

5. Use an experimental shop approach.

The Viking Project was a two-spacecraft mission to Mars, both scheduled for launch in the summer of 1975 on a Titan/Centaur launch vehicle. Each spacecraft included an orbiter and a lander capable of soft-landing on the Martian surface and conducting a series of meteorological, biological, and planetological experiments.

Viking experienced a considerable cost growth, from $364 million estimated in 1968, to $930 million projected in 1975. Factors contributing to the cost growth included: a lack of understanding of the magnitude of the project; use of cost estimates scaled up from the previous Lunar Orbiter project; poor appreciation of the effects of inflation; no reasonable industry cost estimates; and an inability to pinpoint critical technological areas requiring state-of-the-art improvement. It was not clear that additional money during the early phases of the project would have been used to the best advantage because the real problems were not then understood. The project also suffered from insufficient inhouse engineering during its early phases contributing greatly to later problems, especially state-of-the-art improvements.

Angelo Guastaferro, the deputy LaRC project manager, also thought the role of the scientist/principal investigator in all projects should be re-examined. The principal investigator on Viking had no direct responsibility for schedule and cost, and limited responsibility for the performance of the experiment hardware. A consensus was that the scientist should be given the total job, including responsibility for cost, schedule, and performance. The Viking experience showed parametric cost estimates to be totally inadequate.

Recommendations:

1. Realistic cost estimates must be developed prior to large expenditures of project funds.

2. Science definition and scientist participation in instrument development should be managed firmly.

3. Beware of "state-of-the-art" pitfalls.

4. Invest significant early money in hardware development and testing.

5. Assign well-trained contractor management teams to major, critical sub-contractors early.

6. Beware of contractor cost estimates for subcontractor work and changes to complete.

7. Maintain a dollar-reserve posture equal to the degree of uncertainty.

8. Have a continuous cost-offset/cost-concern program.

9. Use an aggressive management and flexible staff concept:

   – Assign Tiger Teams, get outside help, use incremental reviews, keep organization dynamic matched to phase of project.

10. Establish management techniques for control, monitoring, evaluating, updating and reporting of cost, including indirect cost.

11. Assign technical/schedule/cost responsibilities for each area of work to a technical manager.

The Delta launch vehicle project was at this time involved in an adaptation of an inherited design. The vehicles were built in a limited mass production operation. The GSFC project management team was primarily concerned with providing its customers a highly reliable launch system at a reasonable cost. A major concern of the project was determining the proper balance between achieving greater reliability and performance and maintaining a competitive price.

Bill Schindler, the project manager, speculated that in selecting reliability goals for launch vehicles, consideration must be given to both launch vehicle and spacecraft cost. In general, for non-redundant vehicles, reliability levels greater than 90 percent are achievable only at considerable costs, and for reliability goals above 95 percent the cost may well become prohibitive. He felt that in attempting to assess launch vehicle cost versus reliability, spacecraft cost must be considered; that is, a higher spacecraft cost justifies more effort on launch vehicle reliability.

Delta launch vehicle failures had been determined to be about equally divided among electrical, mechanical, structural and ordnance (including solids) subsystems.

The project manager stated there was a large quantity of data on projects that varied greatly in their approach to reliability, from "low-cost" projects such as Delta, Scout, and Explorers, to "high-cost" projects such as Saturn, Apollo, and Viking. He suggested a study to determine whether a quantitative relationship could be established between dollars invested and achieved reliability. He identified the following as cost drivers during development of major configuration change: component qualifications, systems integration and compatibility testing, and formal system qualification. During the operational phase he identified production acceptance testing, requalification requirements, systems acceptance testing, the amount of field rework/modification permitted, and the flight readiness review process.

Following Schindler, Lee Belew, MSFC project manager, presented Skylab, this nation's first space station. Skylab made maximum use of existing launch vehicles, spacecraft, hardware, facilities, and equipment. The management experience from past programs and the ongoing Apollo Program was fully utilized. With the Apollo spacecraft attached, Skylab was 118 feet long, weighed approximately 100 tons, and cost approximately $2.5 billion. Skylab was equipped with solar telescopes, Earth sensors, and equipment for space manufacturing. Skylab was launched on a Saturn V launch vehicle, the Apollo spacecraft on a Saturn 1B. Program emphasis was on obtaining biomedical, Earth applications, and scientific data. More than 100 different experiments were conducted, involving many scientists and principal investigators.

The design, development, test and checkout, launch and mission operations were carried out using essentially the same team. The principal investigators, the scientists, and the crew actively participated in all the above activities.

Skylab. *Photo courtesy of NASA*

The following program cost drivers were identified: (1) Skylab was coupled to Apollo. Apollo supported the program relative to common hardware. Skylab launches were in series after completion of the Apollo program. (2) Crew safety, mission objectives and program requirements dictated a design with considerable redundancy. (3) Skylab was a manned, one-of-a-kind, national commitment.

Recommendations:

1. Make authority delegations known throughout the project organization.

2. The most cost-effective path is to use proven hardware and existing components or systems.

3. Search for available hardware items.

4. A deliberate matching of management skills is recommended when the working relationship involves multiple Centers.

5. If cost is to be the controlling factor, establish it early in project planning.

6. Ensure a strong inhouse systems engineering and integration activity throughout the program.

The Pioneer-Venus Project managed by Skip Nunamaker of ARC consisted of two launches to the planet Venus scheduled for 1978. The orbiter was to be launched first, followed by the probe. Venus encounter was planned to occur in December 1978, for both the orbiter and probe. The probe was designed to enter the Venusian atmosphere and transmit atmospheric data until it impacted the surface. The cost was estimated at $173 million for a six-year period covering fiscal years 1975-1980.

Hughes Aircraft Company was the spacecraft systems contractor for both orbiter and probe. The decision to change launch vehicles from Thor/Delta to Atlas/Centaur allowed much more flexibility in the spacecraft probe design, and contributed to containing costs. Also, the contractor was instructed to plan spare or vacant time in the schedule following each major test. This permitted resolution of test anomalies without impacting other scheduled activities.

Recommendations:

1. Keep mission objectives specific.

2. All mission and spacecraft specifications should be prepared inhouse and given to the contractor, not the other way around.

3. Spend time studying and engineering the proposed mission prior to project start. This will pay big dividends later, especially in cost estimating.

4. Provide pre-project approval funds for ordering parts. Parts availability and long lead times are big cost items and difficult to control.

The High Energy Astronomy Observatory (HEAO) consisted of three low Earth orbit missions whose objectives were X-ray, gamma-ray, and cosmic ray astronomy. The spacecraft was built by Thompson-Ramo-Woolridge (TRW).

The project manager Fred Speer emphasized the thoroughness of definition that preceded the hardware phase and the participation of MSFC engineering in all essential designs. A very high percentage of components and subsystems were off-the-shelf designs, obviating the need for full qualification testing. Major cost savings were accomplished by accepting the protoflight concept on all instruments and the spacecraft. HEAO instruments were constrained to allow for substantial initial design margins in weight, power and volume. Early cost ceilings were established on all instruments, and descoping was performed on those that exceeded ceilings.

A project cost-benefit practice involved the use of common electronic piece part suppliers for both the spacecraft and science experiments. Obtaining piece parts is a major problem for all programs, but especially for experimenters.

Recommendations:

1. Refine and reduce programmatic requirements and concentrate on specific technical requirements.

2. Use value engineering and motivate contractors by sharing savings from cost reductions.

3. Establish firm budget ceilings for each program element.

4. Adopt modular payload mode with options for deletion.

5. Ensure that experiments are manufactured by qualified hardware contractors.

6. Encourage commonality and standardization.

7. Use a design-to-cost approach.

8. Establish adequate contingency funds.

The sounding rockets presentation concentrated on the launch vehicle aspect of project management and did not cover payloads or spacecraft. Sounding rockets are inherently low in cost and take a different management approach to cost control because a large portion of the launch vehicle and payload hardware is recoverable and can be refurbished and reflown. The refly option reduces cost to the point where total reliability is not the concern it would be for a larger, more expensive mission. When failures occur, they are handled in a less formal atmosphere, and the resulting change in hardware or procedures is usually minimal. The project manager emphasized, however, that a launch is never allowed to proceed with a known defect in either the launch vehicle or payload.

Recommendations:

1. Establish better flight program definition.

2. Improve the procurement process for standard hardware by lessening the time and eliminating paperwork and improve NASA cost accounting procedures.

3. Establish methods of evaluating scientific value of the mission and trade against the cost to conduct the project.

Workshop Summary

During a general discussion period, the project managers were asked: What would motivate you to bring in a low cost project? One manager responded: "Not a damn thing. I held cost and schedule. I did, I think, a great job. Then Headquarters comes back and cancels the program." Another project manager said: "No project managers I know were called in and given a promotion or demotion for cost. How many do you know that haven't moved up because of their cost performance?" A third replied quite simply: "I was motivated by the limit on the dollars we could get."

A brief executive session was held to react to the workshop experience. Two issues emerged. First, it was evident that each project manager was avidly interested in the other presentations, wanting to capitalize on the experience of colleagues. How best to capture this experience and make it available to others was the primary concern of the panel.

Second, the group recognized that NASA had no formal training program for its project managers. The experiences discussed at the workshop could be passed on to future managers, but other experiences were being lost. The need for formal training was mentioned more than once.

About a week later, each participant was asked to evaluate the workshop. To a person, the Project Management Workshop was considered a huge success. While one project manager claimed "some people gamed it, protected their rear-ends," another called the sessions "open and frank discussions." One project manager flatly asserted: "The Agency is not yet philosophically ready for low costs." However, each was pleased with the workshop and seemed especially glad to have

Gilruth and Silverstein in attendance. Most of the project managers wanted the Agency to hold such workshops on a regular basis.

On March 11, Metzger wrote Gilruth with his impressions of the first workshop. Metzger's comments were of great interest to the LCSO since COMSAT was frequently held out as a low cost paradigm. He thought the Mariner Venus/Mercury project was successful for a number of reasons:

Gene Giberson pointed out that the fundamental mission objectives were never a problem. All concerned always knew what was to be done from an overall viewpoint.

They spent about a year on "pre-project" planning for the experiments. They spent another year with JPL spacecraft design people going into a detailed paper study on how a satellite could be configured to accommodate the desired experiments. . . .

He claimed the design selected "was able to use previous hardware to a major extent." I consider this approach to be the hallmark of our successful satellite programs.

They held industry briefings and received comments from individual contractors on all aspects of the proposed specs and contract, prior to letting out the formal RFP. He indicated "the comments received had a significant effect on what JPL did."

They selected a contractor (Boeing) who built a somewhat similar spacecraft, the Lunar Orbiter. Their internal budgeting included a contingency.

All of the experimenters had previous experience with spacecraft equipment except one who asked for JPL help in specifications and test procedures and used these . . .

During the negotiations, JPL and the contractor worked up a mutual understanding of the specifications and as a result undertook to modify the specifications to some extent. We at COMSAT have always used this approach and found it essential in this highly technological field. We are just not smart enough to think we can write a specification and expect it to go into the contractual stage and then production unchanged.

They had close cooperation between the cognizant engineers at Boeing and their counterparts at JPL throughout the project, at an engineer-to-engineer level . . .

The contractor was able to draw from a 30,000-man organization as required . . .

110

All in all, the JPL people certainly followed all the rules in designing and building a satellite, and this project could well be used as a model on what should be done for similar projects in the future.

Sphinx—Here we have quite a different type of satellite from the MVM described above. Here was a highly specialized experiment to be incorporated into a new satellite, both to be built in a year and a half. The basic mistake in the Lewis approach was to go to Boeing for the job. That's like using a 155 mm howitzer instead of a .30/.30 rifle. . . . The conclusion to be drawn from this project is that highly specialized experiments involving a simple low-cost spacecraft (in this case about $1 million) should be designed by the people directly involved, augmented by a project manager but one who recognizes that he is in charge of a single experimental satellite project rather than say, a project to build 1,000 Minute Men missiles.

Delta Launch Vehicle—I have always considered the Delta one of the most successful projects in our space history, with a 90% success record in over a hundred launches. It has grown from an 85-pound synchronous orbit capability in 1965 to a 1,000-pound capability in 1975. This growth has been accomplished in all cases by using engines for the various stages from other projects, which is in accordance with the "rule" to use existing hardware wherever possible. Another point which certainly contributed was brought out by a statement by Bill Schindler, that "they never tried to shave weight to the last pound." Still another reason for its success has been the use of practically the same team throughout the 15 years of the Delta's existence.

I would like to emphasize the importance of using previously flown subsystems, components and techniques from other successful satellites in order to provide the best insurance against failure. This approach has been proven in the successful Delta program, the JPL satellites, the RCA meteorological satellites (both classified and unclassified), and the Hughes communications satellites. Failures in these programs have been primarily due to using new designs rather than due to "normal" failure of a component . . .

Viking—I believe there were several basic rules which were not followed in this project. The first was giving the project to a group (contractor) which, in my understanding, had never previously designed a satellite. Their paper design of the Viking resulted in an estimate of $364 million which has now increased to about $990 million, a factor of almost 3 times.

A second mistake was in not doing what JPL had done in the MVM; namely, had talks with contractors prior to going out on contract. Recall

Capsule Tests European Technology, Management

ESA's Atmospheric Re-entry Demonstrator (ARD), which will be the first re-entry capsule made in Europe, is a test bed for cutting ESA red tape by giving industry wide leeway within a budget etched in stone . . .

Keeping within ARD's tight budget, Aerospatiale is using equipment already available rather than designing program-specific components.

by Peter B. DeSelding in Space News— March 18-24, 1996, p. 8.

that as a result of JPL's briefing talks, they had "significant" changes in their specifications. This undoubtedly would have happened in the case of the Viking since the first proposals were about twice as high as Langley's estimate. I believe they made a third mistake by not going with a contractor who had previously done similar work. In this case the project was to build a Mars lander, and one would have thought that the experience of Hughes and their lunar lander (Surveyor) or in Grumman and their Apollo lander, might have provided useful background.

Delta Launch Vehicle.   *Photo courtesy of NASA*

General Comments— I believe that the idea of having project managers at NASA get together to compare notes is an excellent one. I am not sure as to the best mechanism to accomplish this; perhaps several should be tried to find out the one or ones which work best. For example, it might be useful for a number of current project managers to visit one of the Centers (or a contractor's plant) the day before the monthly program meeting takes place; be briefed by the NASA project manager of that particular project on its background and current status; and then have the visitors attend the next day's meeting as observers. This can be rotated so that each has a chance to see the other in action.

I agree with Abe that writing up a series of case studies for both successful and unsuccessful projects may be useful and is worth trying. While this is useful background material, a listing of cases and rules isn't sufficient since the difficult problem is, of course, selecting which rules to apply to a specific project.[3]

Second Workshop

The second workshop was held on June 3-4, 1975, and again, the National Academy of Sciences hosted the session.[4] In order to provide more time for discussion, the number of projects to be covered was reduced from nine to six.

The Projects

The first project discussed was the Orbiting Solar Observatory I (OSO-I). The OSO Project, dating back to 1959, consisted of a series of seven satellites prior to OSO-I. Ball Brothers had built all previous spacecraft; however, due to major changes, the I, J, and K spacecraft were competed, with the Hughes Aircraft Company the winner.

The primary objective of the OSO-I mission was to investigate the lower corona of the Sun, the chromosphere, and the interface in the ultraviolet spectral region, to better understand the transport of energy from the photosphere into the corona, with the secondary objective of studying solar X-rays and the Earth-Sun relationships and the background component of cosmic X-rays. OSO-I consisted of one mission using a 2,340-pound spacecraft with corresponding 827 pound payload carrying eight experiments. Orbital altitude was to be 320 miles circular at 33 inclination. Delta was the launch vehicle.

OSO-I was estimated to cost $58 million and weighed in at 2,340 pounds. Robert Pickard, the GSFC project manager, identified these cost drivers: control system complexity and precision, stored command processor, development of special integrated circuits, inability of Government to maintain funding when needed, and experimenters building their hardware. Elements of cost control exercised by the project were: freezing the design, descoping, establishing cost ceilings on experiments and spacecraft, use of financial reporting, weekly manpower tracking, and frequent contractor reviews.

Recommendations:

1. Build experimenters' hardware to their specifications.

2. Use standard components and subsystems.

3. Establish adequate funding contingencies.

4. Freeze designs early and do not overdesign.

5. Make subsystem engineers fully responsible for cost, schedule, and performance.

6. Believe the cost model, not the proposal.

The Small Astronomy Satellite (SAS) project consisted of three spacecraft: SAS-1, launched December 1970; SAS-2, launched November 1972; and SAS-3, launched May 1975. The philosophy of the SAS program was to build a basic spacecraft and attach an experiment to it. This SAS-3 mission objective was to survey the celestial sphere for sources radiating in the X-ray, gamma-ray, ultraviolet, and other spectral regions, both inside and outside of our galaxy. The spacecraft weighed approximately 262 pounds with a 169-pound experiment package. The orbit was to be 300-mile circular equatorial. The launch vehicle was a Scout. The main elements of SAS management were one leader who makes the final decisions and close teamwork by a small project team of high quality. Marjorie Townsend, the project manager, added these additional elements: conservative design concepts, control of manpower, parallel design on critical items, careful selection of parts and materials, good communications with contractors, selective testing to minimize cost, the ability to predict problems, and good schedule control.

Recommendations:

1. Start experiment development before spacecraft development.

2. Buy long lead time items early.

3. Implement configuration management after design phase and control changes.

4. Have business people on project team to control costs and predict overruns.

5. Work closely with contractor.

6. Use existing designs where practicable, but don't forcefit old designs.

The next project was the High Energy Astronomy Observatory (HEAO) Experiment Package. The Marshall Space Flight Center (MSFC) was responsible for the management of the HEAO Project with the GSFC providing two experiments. These experiments, a cosmic X-ray and a solid state spectrometer, were built inhouse. The GSFC project office headed by Ron Browning managed the hardware development and was the single point of contact with MSFC on all matters related to the experiments. The goals of the project office were to accomplish the program on schedule within cost, incorporating maximum hardware commonality between experiments, and eliminating unnecessary redundancies in the design of each experiment.

Recommendations:

1. Establish necessary resources early to meet other Center requirements.

2. Establish understanding at the beginning between Centers as to how project will be managed and controlled.

3. Keep spacecraft development in parallel with experiment development.

4. Prepare a well-defined statement of work and specifications.

The Hawkeye/Neutral Point Explorer Project was a 68-pound Scout-launched spacecraft built by the University of Iowa. The mission objectives were to study the topology of the magnetic field at large radial distances over Earth's North Polar Cap and the interaction of the solar winds with the geomagnetic field. The university did an excellent job; the project came in ahead of schedule and under cost. Unique features of this project included a one-year Phase B study effort prior to project approval, and an understanding with the university that funds for overrun were simply not available. An overriding consideration was to meet the optimum launch time (April through June).

The Dual Air Density (DAD) Explorer Project consisted of two satellites to be launched into coplanar polar orbits by a single Scout launch vehicle. Each sphere contained a mass spectrometer furnished by the University of Minnesota. The objective of the DAD mission was to study the vertical structure of the density, composition and temperatures of the upper atmosphere.

The LaRC project manager for both Explorer class projects, Claude Coffee, mentioned that cost limitations resulted in an 11-month schedule slip. The greatest impact of the delay was on inhouse personnel, resulting in increased institutional management charges. Problem solving using inhouse manpower also tended to increase project costs.

Problems encountered included the lack of early engineering support, Viking problems impacting project personnel at various times, and the procuring of high-quality materials for the inflatable satellite.

Recommendations:

1. Extensive Phase B studies should be performed for both the inhouse and contracted effort. This requires resources of both people and funds.

2. Develop "baseline" design specifications and interfaces early.

3. Use fixed price subcontracts.

4. Be cost conscious and impress this on the contractors.

5. Avoid research and development after project start.

6. Establish realistic schedule.

7. Develop good relationships between project and contractor teams.

The original Centaur upper stage designed in the late 1950's needed updating. A development contract was awarded to Convair in September 1969 to design, develop, manufacture and deliver one improved Centaur (D1) upper stage vehicle. Included in the contract were special test equipment, a ground station, tooling, a two-burn configuration, and flight software. The basic contract of $24 million included a cost plus incentive fee/award fee, unique for its time. The contract was later increased to $50 million to provide a Titan (D1T) proof flight vehicle and a D1A vehicle for Pioneer G.

The program was completed 4.8 percent under cost, the end items were delivered on schedule, the D1A vehicle met all objectives and, although the D1T vehicle proof flight was terminated by an equipment failure, the validity of all the new developments was demonstrated. Elements of program management included adequate definition, "Development Shop" organization, detail definition of cost, and a contract structure providing motivation and flexibility.

The Development Shop organization, an outgrowth of the World War II "Skunk Works" organization, simplified procedures and paperwork, and decreased formal documentation and reports. Segregation of program activities in controlled plant areas resulted in the direct association of design engineers with fabrication, assembly, and test personnel, and a simplified drawing system. The contractor program manager was given overall responsibilities for technical, schedule and financial aspects, resulting in a highly motivated government-contractor team with excellent communications. Government and contractor used identical schedules, financial data and technical requirements. Contractor engineers were designated total responsibility for technical, schedule and financial performance. Successful program management elements included thoroughly understood tasks, cost definition based upon realistic goals with detailed backup rationale, motivating contract features, proper program management organization at NASA and contractor, and appropriate management systems and tools.

Studies revealed the need for a Centaur standard shroud. A development contract was awarded to the Lockheed Missile and Space Company, Inc. Lockheed had extensive experience in building similar large shrouds and had a proven separation system. A cost plus incentive fee/award fee contract was again used; however, this contract experienced a large cost overrun and cost growth. The major reasons for the cost growth were the Lockheed design which caused extensive interface revisions, the contractor scrapping the "Development Shop" approach, extensive personnel turnover and overhead and labor rate increases caused by reduced business volume.

All hardware was delivered on time and all major milestones were met. The shroud successfully passed all qualification tests with relative ease and performed flawlessly on proof flight and Helios-A launches.

Recommendations:

1. Contract should not be started with major inadequacies in the work statements.

2. A "Development Shop" contractor organization is mandatory to control costs on contracts with a potential for engineering or schedule changes.

3. Contractor top level management attention and authority are vital in controlling expenditures of contractor organizations not under direct control of project office.

4. Sound interface definition between contractors is often the critical factor in controlling overall project costs, and is worthy of the utmost attention of contractor and NASA upper management.

The final project discussed was Mariner Mars 71, managed by the Jet Propulsion Laboratory (JPL). The primary mission objective was to study the dynamic characteristics and to provide broad area observations of the planet Mars from Martian orbit. The project was formulated in the face of a threat that no new planetary programs would be approved unless attractive low cost systems could be provided. During this period both the Mars 71 Probe and the Voyager projects had been canceled. The spacecraft weighed 2,266 pounds with an instrument package of 151 pounds.

The 1971 launch date provided a unique opportunity for orbiting the planet in that it was the lowest energy year in the 15-year cycle, and the cost effective Atlas Centaur launch vehicle could be used. The original approach was to use the Mariner Mars 69 science payload with no significant modifications. However, this approach was subsequently changed to include additional instrumentation, modifications to the Mariner Mars 69 instruments, and broader involvement of science investigators. These changes resulted in a cost increase from the initial estimate of $93 million to $106.3 million. JPL managed the project in the subsystem contracting mode.

Cost drivers included inflation and the following three mission scope changes: science experiments, mission operations, and science data analysis expansion. These costs were partially offset by deleting a third spacecraft.

Recommendations:

1. Initial cost estimates should include an allowance for inflation, and some contingency to cover costs of unforeseeable problems.

2. A definitive statement of science payload requirements, together with an estimate of instrument development costs and their effects on spacecraft costs, is needed.

3. Standardize, wherever practical, on designs, components, and test procedures and undertake block buys of identical hardware subsystems.

Workshop Summary

This ended the second workshop. These programs and projects—ranging in cost from $1 million to $2.5 billion—demonstrated not only the vast diversity of NASA activities but also the wide differences of opinion and strong, independent thinking on the part of NASA program and project managers.

What were the results of the two workshops? Metzger's notes are a good guide for the first workshop, and at the conclusion of the second workshop Pete Taylor, a staff member of the ASEB who attended both sessions, wrote his impressions to an ASEB member. "The Second Joint Project Managers Workshop was considered a success by all attendees," he said. Then Taylor listed several needs that he saw emerging from the workshops:

- Improved recognition by senior management of those project managers who complete their project within cost or who exceed performance, and milestone goals using low cost management techniques.

- The difficulties of emphasizing low cost if senior managers materially change program requirements once the program is underway.

- The increasing need for a "piece parts" bulk purchasing and test group within NASA who could properly burn in and qualify parts such as resistors, capacitors, diodes, etc., to satisfy all Agency requirements. (Recent failures, resulting in project delays and or failures running into the hundreds of thousands to millions of dollars, have been traced to unexpected and premature failure of such items.)

- The increased need for project manager training in the areas of planning, budgeting and finance prior to becoming a project officer.

- The vital importance of accomplishing a detailed definition study in which NASA and contractor personnel jointly work out and agree to all management, cost, scheduling and technical details of the proposed contract package before either side becomes a party to the final contract. . . .

- The urgent need to develop a generalized basic cost model which can be used by the prospective project manager in initial efforts to estimate project costs in detail.

- Preparation and utilization of an "approved list" of piece parts, components, subassemblies and subsystems which can be used by future managers to hold down costs. . . .

- Support for a study group to devise and develop new low-cost management techniques necessary to hold down the cost of space experiments during the era of the Shuttle . . .

- The necessity for a highly competent individual (or team depending on the size of the contract) representing the project manager at the contractor's plant to seek out probable problem areas before they can become serious.[5]

Almost to a person the participants thought the workshops to be of immense value and that similar meetings should be held on a regular basis. Cost was discussed, not to the extent the sponsors had hoped, but it was now on the table as a legitimate management concern. If the participants had not become full converts to the relatively new cost reduction drive, they at least better understood the motivation behind it and the need for it.

As to the specific cost cutting recommendations, a number of clusters could be identified. One cluster centered on the need to motivate both NASA and contractors to perform in a low cost manner, and another on the need to assemble the best possible team.

George Low, who frequently began his low cost speeches with "Don't reinvent the wheel," was pleased that most of the participants recommended, in one way or another, the use of flight-proven equipment, off-the-shelf hardware, proven parts and common systems, and previous designs.

The need for good definition before the start of the final design and fabrication phase was mentioned a number of times. This included the need to get adequate engineering support, both inhouse and contractor, early in the definition phase.

This issue subsequently appears in numerous studies on how to improve NASA project management.

A number of cost control processes were recommended:

• Fixed cost, variable scope approach

• Cost-trade analyses

• Design to cost

• Value engineering

• Cost offset processes

Other suggestions included:

• No research and development in the development phases

• Use of experimental shop concepts

• Simple interfaces and controls

• Use of fixed price contacts where possible

• Restraint in the size of the project staff

• Use of specific and detailed RFPs

Third Workshop

A third workshop was held on September 17-18, 1975, at the McDonnell Douglas Astronautics Company, Huntington Beach, California. This workshop gathered the industry managers for seven of the projects covered in the previous NASA-oriented workshops. The projects included the Atmospheric Explorer, the Delta Launch Vehicle, HEAO, MVM73, Pioneer-Venus, Skylab and Viking. No minutes of this meeting exist but a number of participants did exchange letters on their impressions. One such correspondence was again between Sid Metzger and Bob Gilruth.

On October 7, 1975, Metzger wrote Gilruth of his impressions of all three workshops:

Overall I believe the meetings were very successful if they did nothing else than bring these people together and give them a chance to listen to their counterparts talk of their experiences with other projects. This was true not only for the NASA group but also for the industry group. I spoke to a number of managers and they all felt that they had learned something

120

worthwhile from the meetings. The conclusions that I drew from these meetings are summarized below:

. . . The contractor's approach and organization should be matched to the project, i.e., don't give a "Sphinx" satellite to Boeing, just as it would be a disaster to give a "Viking" project to Van Allen at Iowa State . . .

The concept of a training course for NASA fledgling project managers has been thoroughly discussed in the past and agreed to by all. It is put down here again just for completeness. The training session should emphasize that each project is different from other projects and requires a different approach. This is important because there is always a tendency to devise a standard procedure which can be used for all problems. With NASA satellites differing as widely as a Hawkeye to a Viking, clearly a single magic approach would not apply. In addition, I would believe a meeting of project managers about every three years could usefully be held in order for them to compare notes as they did at our meeting.

The problems of a centralized purchasing system are so many and so great that I shudder at the thought of proposing this. However, almost every speaker mentioned problems in this area (of components) and I would therefore recommend that the subject be at least looked into as an experiment . . .

All of the usual generalities and clichés concerning project management were voiced at these meetings but the following items were those brought up the largest number of times, and therefore might be of use to NASA in planning their future projects.

a) Let the work statements pin down in more detail than heretofore just what is to be accomplished in a given project. This could best be done after discussions among the NASA team, the experimenters, and industry.

b) If new critical subsystems are needed, work should be started on that prior to permitting its use in the final spacecraft.

c) Give the contractor a chance to do the job assigned before the NASA project manager gets deeply involved.

All too often, especially in the early days of NASA, project managers learned on the job. Experience is a good teacher, but little if any learning between other project teams takes place. There was no logical reason why NASA managers had to learn only from their mistakes when they could learn from the mistakes—as well as successes—of others. The workshop series was a step in the right direction. Here they produced lists and lists of reminders and suggestions from program and project managers, many of whom had gone on to lead bigger and better programs within the Agency and in industry. By comparing and contrasting methods and techniques, it became clear that no one way is best to plan a program and manage it, some ways are certainly better than others, and some lessons are learned, never to be repeated or forgotten. At the end of the workshops, the participants and sponsors tried to synthesize the experience and follow up with action items.

Vince Johnson, with the full endorsement of the sponsors and participants, forwarded a set of recommendations to the Deputy Administrator. He wrote in part:

I feel that the workshops turned out to be interesting meetings, of value to the participating project managers, the ASEB members, and potentially at least, to NASA management.

The workshops produced an abundance of recommendations from the project managers. Enclosed are a dozen recommendations which Frank Hoban and I feel are worthy of particular attention.

There is still some question in the minds of many project managers as to the reality of NASA management's commitment to "low cost." They are all aware of the stated goals, but there is considerable doubt that management really means that low cost is of comparable importance with success. This presents a very real dilemma to management, since awards and promotions have logically followed successes, and it would be pretty difficult to reward a "low cost failure," thus the project managers really feel that success is all important and high costs will be forgiven. On the other hand, in reflecting on the discussions at these meetings, it is apparent to me that the project managers are, in fact, emphasizing costs much more than in the past, and progress toward low cost approaches is really occurring.

The serious concerns, expressed particularly by Gilruth and Silverstein, on the loss of accumulated project management experience from NASA are well-founded . . . However, I feel that the general quality of NASA's project managers is still very good, and the project managers at the meetings were capable people who demonstrated that not all of the good project managers have been promoted or retired!

General Recommendations:

1. Training for Project Personnel

    While NASA project managers should probably continue to come from the ranks of successful deputy PMs and system managers, I believe that a need does exist for more formal training for project personnel. At least two approaches are suggested:

a. Sample the Defense Systems Management School

b. Set up a shorter (two to three week) training experience directly oriented toward NASA projects as a pilot effort. This will require a lot of work, but it should be a serious and good effort or not done at all.

2. Periodic Meetings of Project Managers

I feel that periodic (perhaps annual) meetings of NASA project managers would be valuable both to NASA management and to the project managers themselves. . . .

3. "Lessons Learned" Reports

A general concern was expressed by many at the workshops that the special knowledge gained by NASA personnel in the management of large complex systems is in the process of being lost, particularly with the loss of key people. Training and project manager meetings would help here, but it was also felt that a good "Lessons Learned" report for each project would be valuable . . . It must be done by the key project personnel themselves, and should probably evolve throughout the life of the project.

4. Independent Cost Review Teams

The problem of obtaining accurate early cost estimates for projects is universal—and truly difficult. One useful tool can be an early review of each project by an "independent" team of both business and technical people before approval of new starts, and perhaps later in the project's life.

5. Enhancing "Low Cost" Motivation

A recommendation is clearly needed here, since it was apparent that the project managers at the Workshop were not really strongly motivated toward low cost approaches in the same way in which they were motivated toward mission success and scientific and engineering excellence. Unfortunately, I do not have a good recommendation to make . . . I believe the attitude toward low cost approaches is changing, but slowly.

6. Agency Piece Parts Program

The problem of obtaining reliable electronic piece parts at a reasonable cost was raised many times, and appears to be getting worse. I feel that serious consideration should be given to setting up an Agencywide electronic piece part program at a single Center which would qualify and procure parts for NASA launch vehicles and spacecraft.

7. Reliability Study

    A suggestion was made that NASA should now have enough historical data to provide a quantitative assessment of the effect of reliability efforts expended on projects. That is, one could review the detailed results of projects like Saturn versus Delta, Mariner or OAO versus Explorers, etc., and attempt to quantify the value of the added specific efforts expended on reliability in those projects. I am not personally optimistic that such a study would yield meaningful results but would recommend that the reliability group look at the question in a bit more depth.

8. Research and Development in Flight Projects Versus "Enabling Technology" under SR&T

    A major cost driver in flight projects is the research and development of new technology frequently required—and sometimes not anticipated—in flight projects. There was general agreement that ideally such efforts should be carried out under Supporting Research and Technology (SR&T) prior to commitment of "new" technology to a flight project, but with recognition that this does not, and cannot, happen in all cases. Again, there was no simple recommendation for a "solution" here, but it does appear that more emphasis on identification of new technology requirements before project approval is called for, as well as better communications between project planners and SR&T people.

9. Pre-Project Approval Buys, and Block Buys

    Several project managers felt that substantial cost reductions could be achieved if they were able to buy long lead time parts or subsystems prior to formal project approval, particularly if block buys of major subsystems could be made . . .

10. Funding Contingencies

    . . . The Agency should do a better job of providing—and protecting—funding contingencies. There is still a reluctance to identify contingency funds due to the fear, frequently justified, that they will be taken away. The resultant lack of contingency funds could thus adversely affect the project's schedule and final costs.

11. Cost at Completion Versus Cost Per Fiscal Year (FY)

Several project managers reported that they had encountered problems due to shortages of funds in a given fiscal year as opposed to cost growth at completion, and that FY shortages caused slips in schedule and total cost growth. . . . It can be a real problem where the government or contractor has not properly identified cost requirements by FY, or where FY fund availabilities change.

12. Headquarters Role in Project Management

While most project managers at the workshops felt that they had adequate authority to do their jobs, a few felt that the Headquarters Program Office—or NASA top management—exercised far too much control over the details of the project, with an adverse effect upon efficiency and costs. While the project managers understood the Headquarters interest in troubled projects, they were unanimous in feeling that Headquarters frequently went too far in the matter of detailed control over projects. I concur in this opinion, and feel that Headquarters personnel at all levels should leave the details of project management to the Field Centers and project managers, with, of course, appropriate review and general guidance, and reward or punishment as indicated. I recognize that there is a great deal of judgment involved on both sides here, but a more uniform policy leaning toward less Headquarters involvement in details seems indicated.[6]

George Low approved each of Johnson's recommendations and asked Gray to see that they be implemented.[7] However, little progress had been made before Low left the Agency.

ASEB Report

The ASEB ad hoc committee summarized their participation and recommendations in a draft report entitled "Management of Space Program Costs." The report was not officially submitted to NASA nor does it appear on any listing of ASEB reports. This does not mean that NASA management did not review the report, it simply means the report had no formal standing and therefore did not require an official NASA response.

The report recommended that "a short but intensive training program be provided to prospective project managers" and went on to enumerate the reasons. The report concluded:

In summary, experience has shown that practical training programs for potentially successful managers should emphasize brief but intensive exposure to the suggestions and techniques used by:

1. Successful space project managers;

2. Staff specialists responsible for successful functioning of space project support needs at the Headquarters NASA management level;

3. Key specialists essential to the proper functioning of a project at the Center level;

4. A typical contractor working on a typical NASA space project.

With the guidance provided by such individuals currently and successfully involved in the management process of space projects, the project manager trainee will not only be able to make more effective and realistic use of the tools of the trade, but will be enabled to meet the goals of performance, cost reduction and schedule completion on the project for which he is given management responsibility.

The report also endorsed that NASA project managers "be provided an annual opportunity to meet in a workshop forum to present and discuss their successes, mishaps and application of new management techniques in an effort to ensure more effective project management across the board within NASA." The report went on:

As a potential source of important cost reductions it is suggested that each step involved in the management of a space project from its initial conception to its final successful performance be reexamined with a view to eliminating or modifying the currently accepted requirements, procedures, documentation, etc. that no longer contribute to the cost effective management of a space program or to the achievement of its specified objectives.

The report continued:

To date the Low Cost Systems Office staff in NASA Headquarters has identified, described and focused attention on many ways to achieve cost reductions within the limits of sound management practices. The Committee commends this activity and urges that it be continued. In addition to studies now underway, the Committee suggests that an increasing emphasis be placed on risk and cost tradeoffs involved in the elimination or simplification of procedures and techniques established many years ago which, although considered essential in the traditional sense, may in fact no longer be necessary with the advent of newer technology and management techniques as well as improved instrumentation and testing techniques . . .

NASA should study the cost benefits resulting from the establishment of a centralized purchasing group responsible for the purchase, qualification and issue of electronic components, subsystems and other hardware commonly used in launch vehicles and space experiments, and that the system of incentives and rewards for successful NASA project managers and staff personnel be reviewed and improved.

During the course of both Project Manager Workshop Committee meetings, project managers as a whole indicated that improved visibility within NASA and industry for successful accomplishments as project managers would be highly desirable to improve the morale of this critically important management group. In addition to full use of the limited system of rewards available within the Civil Service system and NASA management procedures, senior managers should further improve the visibility of those individuals who successfully achieve significant cost savings in their work.[8]

The report concluded with a suggestion "that the standard cost models evolved by those Centers and major contractors involved over the years in NASA space activities be used as a basis for a comparative cost study." These conclusions remained in draft form and were never officially sent to NASA.

Years later, when the Low Cost Systems Office became the Low Cost Systems Division (LCSD) of the Chief Engineer's office, the division staff reported on the disposition of the Project Management Workshop recommendations, responding to a query by the Chief Engineer. An abbreviated version of this report shows that most of the "lessons learned" from the Project Management Workshops would have to be learned all over again, for most of them were ignored or forgotten. The staff reported:

1. Training for Project Personnel—The LCSO, with the cooperation of GSFC, put together a pilot "Project Management Shared Experiences Program," first offered in May of 1976. The program was fully supported by the entire Agency and was judged an outstanding success by the participants. The program was then turned over to the Office of Professional Development and was being offered twice a year at the Wallops Educational Facility. This program was the only effort on the part of the Agency to pass on the "Lessons Learned" by NASA project managers to future generations.

2. Periodic Meetings of Project Managers—This recommendation was never carried out.

3. "Lessons Learned" Reports—This recommendation was never carried out.

4. Independent Cost Review Teams—This recommendation was never carried out.

5. Enhancing "Low Cost" Motivation—A contract was let with Development Research Associates (David Berlew) to look at the broad aspects of this question. (See Chapter 10.)

6. Agency Piece Parts Program—An active Piece Parts Program has been established.

7. Reliability Study—This recommendation was never carried out.

8. R&D in Flight Projects versus "Enabling Technology" Under SR&T—This recommendation was never carried out.

9. Pre-Project Approval Buys and Block Buys—This recommendation was never carried out.

10. Funding Contingencies—This recommendation was never carried out.

11. Cost at Completion versus Cost Per Fiscal Year—This recommendation was never carried out.

12. Headquarters Role in Project Management—The intent of this recommendation was partially covered in the Project Manager Review procedures. (However, before the next budget cycle, Low had left the Agency and only one or perhaps two such reviews were held.)

Thus, few of the recommendations were fully implemented, others languished without sponsors to carry them out, and others were simply ignored.

It was widely recognized by the workshop participants and sponsors that costs could be reduced significantly through the transfer of past experiences to the next generation of NASA managers. How their "Agency memory banks" would take shape was a matter left for another day, but one notion was patently clear as a result of these sessions: NASA needed a project management training program. Initial assumptions on how to tailor such a program for NASA requirements and style included maintenance of a strong inhouse capability, a common institutional environment for management of projects, close monitoring of both government and industry contracts, and new concern for balancing schedule and technical performance with cost, while preserving NASA's historical emphasis on high-level technical success. It was thought that a well-qualified, properly trained project management work force would result in unprecedented cost savings. If the project management team were better trained and motivated, there was no telling what they could accomplish.

Footnotes

[1] ASEB, brochure. National Research Council, Washington, D.C., 1991.

[2] First Project Management Workshop, February 24-25, 1975.

[3] Sidney Metzger, letter to Dr. Robert Gilruth, March 11, 1975.

[4] 2nd Project Management Workshop, June 3-4, 1975.

[5] J. P. Taylor, letter to Frank W. Lehan, June 6, 1975.

[6] Vince Johnson, memorandum to Ed Gray, August 12, 1975.

[7] George Low, memorandum to Ed Gray, September 12, 1975.

[8] ASEB Ad Hoc Committee, Management of Space Program Costs, unpublished report dated November 1975.

Chapter 8

Low Cost Workshop:
A Gathering of Unbelievers

In early 1975, things were not going especially well for the Low Cost Systems Office. True, equipment was being standardized, practices were being scrutinized and project management workshops were underway, but progress was less than Low, Gray or Muinch had expected. The half-hearted support of the Field Centers was of special concern; without the Centers, the standard equipment program was doomed. The LCSO staff reasoned that bold new initiatives were needed to increase the program pace, and one such initiative would be an Agencywide cost reduction conference.

Muinch wrote to Gray on April 4, 1975:

The Low Cost Systems Office has been in existence for approximately two years. The office has, in recent months, made significant progress in standardizing equipment and exploring potential changes to program and business practices. This progress, however, has been made with the wholehearted support of only 2 of the 11 Field Installations. Lack of support from the remaining Installations can be attributed to a variety of reasons, from reluctance to change to considerations of manpower and Installation priorities.

A staff organization of this size cannot force Installations to participate and most certainly cannot force Installations to reorder their priorities. We can only persuade and cajole. Therefore, in order to maintain and expand on the present limited momentum, a new commitment to the objective of reducing Agency cost is in order.[1]

Muinch believed a high-level conference was needed to review the entire LCSO program, surface new ideas and foster a new commitment to Low's vision. But the LCSO staff hoped the main outcome of the conference would be the establishment of a Low Cost Systems Board. The Board would be composed of Deputy Director-level NASA managers from the Headquarters program offices and NASA Field Centers. Its prime purpose would be to ensure the involvement and commitment of the entire Agency to the low cost program. Ed Gray would chair the

Board. Muinch told Gray he expected the Board to be responsible for a variety of activities, but most importantly how a low cost program should be defined, pursued and managed.

Muinch wanted Low to be the star attraction of the conference, inviting the participants and chairing most, if not all, of the sessions. Gray agreed and the wheels were put in motion.

The Low Cost Workshop

The conference, now officially called a workshop, was held on July 8 and 9, 1975, at the Goddard Space Flight Center in Greenbelt, Maryland. By the time the workshop was held, the idea of a Cost Reduction Board was dead. No one outside the LCSO would stand for the formation of yet another NASA committee. Only a few of the invited Deputy Center directors came, and Low attended for only a short time. On the first day, he delivered a stirring address on the need to reduce cost and support the LCSO program, and then left. He returned the second day for the final session. John Naugle the NASA general manager, chaired the rest of the sessions.

Prior to the workshop, the Centers had been asked to provide agenda items. They responded questioning the credibility of the entire low cost effort and its philosophy. They questioned NASA's reliability, quality and testing practices, its procurement practices and the lack of effort devoted to program and business topics. They wanted to discuss how best to develop and supply standard equipment. They wanted to discuss risk, in particular NASA's "no failure" image and how much management was willing to pay to maintain it.

The discussion of risk was very spirited, but no conclusions were reached. When Low returned for the workshop summarization and heard about the risk debate, he requested the LCSO establish a procedure whereby top NASA management and the project manager would define what the project was expected to accomplish, the risk versus cost approach to be taken and the criteria on which the manager would eventually be evaluated. This was to become the flight project review process.

In a few weeks Gray wrote to the attendees.[2] He said it was clear to him the Centers believed they were conducting an aggressive low cost program independent of the LCSO. "It is also apparent that Agencywide functions of the Low Cost Systems Office (LCSO) are not clearly understood," he wrote, admitting the obvious. Using Low's words, he then attempted to define the role of the LCSO as the Agency catalyst on cost reduction. The job, as Gray defined it was:

- Developing standard hardware, software and procedures having multi-mission applications which will reduce the cost to the program offices and/or appropriate to an overall cost savings to NASA. This requires Center participation in the selection, specification and management of those items.

- Searching for opportunities to reduce cost throughout NASA on a continuing basis. Most of these opportunities are located at or are controlled by the Centers.

- Developing a program for motivating and rewarding NASA personnel and contractors for outstanding cost performance.

Gray suggested that to carry out these functions properly would require most Centers to increase the number of employees assigned to support LCSO activities. He asked each Center to develop a yearly plan "which defines its activities to reduce costs and spells out goals to be achieved."

He assigned the following actions to the workshop attendees:

1. Appoint a team to work with the Low Cost Systems Office.

2. Provide him their views on the procedure suggested by Low to require a review between individual project managers and NASA top management, establishing guidelines for the conduct of the project, the cost risk considerations and the criteria for evaluating project success.

3. Put more emphasis on the use of commercially available equipment.

4. Critically examine NASA's procurement practice of requiring the preparation of extensive protest-oriented documentation with the objective of decreasing procurement lead times, thereby reducing costs.

5. Assure the training of potential project managers with the experiences of current and former project managers as part of the training.

6. Agree that a specification review is needed to reduce and consolidate the present inventory of NASA specifications and standards, and support the formation of a full-time team, headed up by a Headquarters designee.

No new support of the LCSO resulted from the workshop or Gray's request. Low's limited participation was viewed as a negative, no Board was established, and the attendees pledged no new help. But two significant outcomes eventually made the exercise worthwhile: a Project Management Review process and a Payload Review process that is still in effect today.

These processes did address the impact of image and risk on NASA costs— two major areas of cost generation that until the retreat were discussed only among trusted associates—a faint indication that a culture change was underway.

Footnotes

[1] George Muinch, memorandum to Ed Gray, April 4, 1975.

[2] Edward Z. Gray, "Low Cost Retreat Minutes" in a memorandum to Center Directors et al., July 22, 1975.

Chapter 9

Shared Experiences: A Renewed Training Effort

The Apollo work force of the early 1960's was arguably among the best ever assembled for any human endeavor. Outstanding individuals from academia, the aerospace industry, DoD and NASA eagerly joined the Apollo team. It was a work force not satisfied just to succeed—it strove to excel. The team was young—the average age of the entire NASA work force was then under 40—but its members had a surprising amount of relevant experience. The combination of a well-defined national goal, enthusiasm, youth, experience and ample resources all contributed to the outstanding success of the Apollo Program.

By the time of the lunar landings late in the decade, the Apollo team showed the first signs of aging. NASA's Office of Manned Space Flight (OMSF) quickly recognized the need to renew its supply of qualified project management personnel. One contribution to this renewal would be a formal project management training program for OMSF personnel, even if it were not Agencywide in scope at this time. Associate Administrator for Manned Space Flight Dale D. Myers wrote to Bob Gilruth, then the Director of the Manned Spacecraft Center in Houston, stating that he wanted a pilot project management training program:

> To provide instruction which will improve the capabilities of selected engineering and management personnel in carrying out their functional assignments in (a) systems engineering and (b) systems management. A secondary but closely-related objective is to develop an appreciation for the importance and practical value of effective communication and related interpersonal and inter-group relationships. . . .[1]

The objectives did not specifically mention cost, however well-managed projects are often, by their very nature, cost-effective especially if they are managed right the first time.

The pilot program, known as the NASA School of Science and Engineering, was designed and delivered by the Harbridge House Corporation of Boston in the spring of 1970. It consisted of four weeks of systems engineering training, held at the Michoud Assembly Facility in New Orleans, and a two-week class in systems

NASA School of Science and Engineering (Pilot) Systems Engineering Course (1970)

The Systems Engineering course was composed of several distinct but related elements:

- Case Studies in Systems Engineering
- Human Behavior in Organizations
- Theme Project
- Tools and Techniques of Systems Engineering
- Computer Skills Program
- Systems Management
- Technical Briefing

management at Airlie House in Warrenton, Virginia. Twenty-five Manned Space Flight employees were the first participants.

Myers had high hopes for his new school. He felt it important that "every organization, including ours, conduct some type of training as an investment in the future." His proposal was to turn out 675 graduates a year at a cost of a million dollars, including course development and pilot programs.

Following the pilot courses, Gilruth wrote to Myers about his fledgling school. Gilruth had understood that the NASA School of Science and Engineering "was established to disseminate the knowledge and experience we have gained from our manned spacecraft programs to personnel within our organizations." However, Gilruth had gotten an earful from a dozen MSC participants in the two pilot courses. "It has apparently evolved into a theoretical management and systems course," he wrote to Myers. "If it is the intent to teach theory, it would probably be to our advantage to send individuals to existing well-established schools."[2] Although this response effectively killed the new NASA School of Science and Engineering, Gilruth's memo did establish the primacy of knowledge and experience over theory in future curricula.

Gilruth's memo to Myers also set the tone for shorter courses in place of the two- and four-week pilot programs. He suggested, instead, "a series of short seminars of two- or three-day duration . . . a combination of theory and practical experience taught by instructors from the Centers, industry, and universities who have had the benefit of practical experience in their fields." With this suggestion, the parameters were set for shorter courses involving veteran NASA project managers as well as outside experts. Quite unintentionally, Gilruth's objections also inhibited the development of any Agencywide project management training for several years.

Meanwhile, in July 1971, the Department of Defense established the Defense Systems Management College at Fort Belvoir, Virginia, and began offering classes to both military officers and civilian contractor personnel involved in systems acquisition management. Some in NASA could only look with envy at the new school's impressive physical plant, its permanent faculty and its support staff.

Low and Gray met with several members of the Aeronautics and Space Engineering Board (ASEB) in the fall of 1974 to review progress on an ad hoc study the Board was conducting to identify project costs and to recommend actions for future savings. During the review, two related issues were discussed. The first was the possibility of the ASEB conducting a workshop with NASA and industry project managers to identify what could be done differently in the management of projects to reduce costs. Second, the ASEB was to recommend to Low "whether or not there should be a project manager's school, and, if so, when in a career should attendance take place?" It was evident that Low had connected trained project management personnel with good cost management.

Subsequently, in the summer of 1975, the first recommendation from the project management workshops was to train the NASA work force. Attendees at the workshops felt strongly that a trained work force was necessary to manage projects in a cost-effective manner. The effort to define such training was greatly helped by suggestions from Angelo Guastaferro, a presenter at the first workshop. In response to a request from Vince Johnson, the workshop moderator, Guastaferro had recommended several topics for improving the project management process, all having training implications. A summary of these recommendations include:

- Training. Establish a project manager's handbook and utilize it in a mandatory training seminar for all key personnel assigned to NASA projects. The course should be run by NASA Headquarters using a case study approach with lectures on financial management, schedules, resource control, procurement, legal, institutional support, organization and management, support services contracts, technical reviews, documentation, configuration control, change management, project management, decision making, reliability and quality, low cost, integration and interface control, multi-Center management and program management.

- Planning. Proposed new projects should be studied sufficiently to establish realistic cost. Technical and cost proposals should be reviewed by an independent group of senior NASA officials. Science definition including user requirements should be established during the proposal stage and not after project approval. NASA should invest in long-lead hardware and software development items.

- Project Implementation. Conduct project manager's council meetings on a quarterly basis in which management techniques and procedures utilized by one project can be evaluated for use by others. Standardize the technical and management review format as to frequency, content and follow-up procedures. Make each project establish a project reserve in both schedule and cost. Project cost reserves should be augmented by cost offsets achieved through efficiencies and program adjustments. Publish a policy and requirements document for each project outlining the methods of controlling configuration, weight, cost and schedule.

- Past Project Activities. Each project should prepare a chronological cost history with an assessment of the cost drivers and subsequent management actions concluding with a section on the specific lessons learned on the project. This history should be distributed to all active and proposed project managers and be an agenda item at one of the NASA project managers' council meetings.[3]

These four recommendations, along with those from the project management workshops and the ASEB report, were to lead to the development of the Project Management Shared Experience Program, NASA's first Agencywide training effort.

137

Shortly after Johnson submitted his recommendations to Low, he and the LCSO staff began to explore the actions necessary to establish an Agencywide project management training program, one that would highlight cost management and include NASA's lessons learned. They examined the training opportunities available through the academic community, the business community, the Department of Defense and other government agencies.

They uncovered a vast array of project management courses available from a diverse set of providers. But taken individually, these courses would require an inordinate amount of time—and the coursework would have to be supplemented by NASA unique project-related instructions, including lessons learned.

They visited the Defense Systems Management College (DSMC). The school's primary mission was to prepare military officers and civilian personnel for assignments in program/project management career fields. The DSMC 20-week Program Management course contained many elements useful to NASA project managers, including systems acquisition, contracting, financial control, logistics and production. One immediate drawback, however, was that only two to four NASA managers could be accommodated at the DSMC each year.

After completing this survey, it was decided that NASA should develop a training program specific to the needs of its work force. The LCSO asked Dr. Michael J. Vaccaro, Associate Deputy Director of the Goddard Space Flight Center, to help. Vaccaro's experience in managing GSFC flight projects was greatly valued, and he had led the development of an effective and realistic project management computer model known as the Goddard Research Engineering Management Experience (GREMEX). This model illustrated the decisions and consequences a project management team would face over the entire life cycle of a flight project, from contractor selection to launch.

Vaccaro began by asking Boston College, the University of Virginia, and California State University in Los Angeles for assistance. Each of these institutions had in-depth knowledge of NASA and its project management system. He asked them to conduct a preliminary study to develop a rationale and approach for the design of a special training program. Each study was to be a detailed approach rather than a final design, and would address at least the following:

- Program Content—Project management tools, key principles, techniques and other factors critical to project management success.

- Program Approach—Rationale and techniques to be used in stimulating and supporting learning to ensure effective transfer of information.

- Final Program Implementation—Estimated resources necessary for the detailed design and development of the final program.

- Other Issues and Concerns—Comments on any other aspects of the problem of preservation and continued adaption of project management capabilities.

Although different approaches and content were proposed, the institutions generally agreed that the program should address team-building, organizational structures, planning and control techniques, cost control, general problem solving and behavioral skills, all within the project management context.

Training Program Proposal

Using this information, Vaccaro, Johnson, the LCSO, and an ad hoc team consisting of Gilruth, Gene Giberson, Lee Belew, and Glynn Lunney developed a "strawman" program for review. The recommendation of this group to proceed with their program was endorsed by Low, who urged Gray to "move out as quickly as possible."

The strawman proposal included elements from the four proposers, the project management workshops, the ASEB draft report and the NASA project management community. Due to management's strong belief that key project personnel could not be spared for more than a week, and that to attract the "right" people the program should be short and intensive, the program was limited to five days. It emphasized those aspects of project management that best defined NASA's approach. Among the initial assumptions about what constituted the NASA style were:

- A strong inhouse capability.

- A common institutional environment for project management.

- In-depth monitoring of contracts, both Government and private sector.

- Historical emphasis on a high level of technical success and a growing concern for balancing cost, schedule and technical performance.

Participants would be exposed to project management practices through techniques that simulated project decisions along with situational exercises, case studies, simulation and symposia using learning situations based on the experiences of project managers.

Participants would be limited to 25 project personnel selected by the NASA Centers. The program would be a NASA-dominated creation with strong inhouse support and a full-time NASA manager.

Training Program Proposal

A. NASA Project Management
   1. Concepts
   2. Policies and procedures
   3. Project plans
   4. Resource acquisition
   5. Project organization structures
B. Specific NASA Projects
   1. Objectives, constraints
   2. Purpose and role of projects
   3. Importance of project constraints
   4. Evaluation and reporting
C. Understanding Procurement and Finances
   1. Contract types
   2. Contract management, negotiations
   3. Elements of cost
D. Understanding Contractor Relationships
   1. Historical performance
   2. Strengths, weaknesses
   3. Corporate objectives
   4. Business environment
E. Tools of Trade
   1. Project plan
   2. Project documents, reports
   3. Project management reporting systems
   4. Technical processes

Final Program Design

In early November 1975, a formal steering group was organized to formulate the final program content and coordinate NASA Field Center support.

Using the preliminary proposal as the baseline and incorporating a steady flow of suggestions from the Field Centers, the steering group developed guidelines for the curriculum content:

• The program should emphasize the "sharing of experiences" by Agency project and functional managers and participants as opposed to tutorial instruction. Participants should benefit most from experiences shared by leaders whom they respect for their accomplishments.

• It should emphasize shared experiences relevant to "real life" tasks and should enable participants to relate technical skills to the art of project management.

The steering committee then developed six major curriculum groupings:

1. Project Management in NASA. Understanding the degree of managerial flexibility available to the project manager as well as the needs of external users and scientists. This will enable the project team members to be more aware of their decision-making options and constraints.

2. Agency Budgeting. Because various aspects of the budget process are interwoven in the project's business, each member of the team needs to understand both the overall budget process and the relationships between each member's work and the budget philosophies of the project.

3. Procurement Process. The many regulations and procedures used to establish the basic agreement between the government and the contractor must be understood by project managers.

4. Understanding Contractor Relationships. By being aware of contractor characteristics and past history, better ways of doing business can be established, providing greater opportunity for success and cost control.

5. Tools of the Project Management Trade. Examine many of the existing tools and their application to unique situations.

6. Project Manager Workshop. This forum, a panel of NASA project managers sharing significant aspects of project management from their own experiences, would conclude the program and it would feature interaction among the participants and the project managers, a technique that had never been used before in NASA training.

The steering group now began to prepare the individual program sessions. Further contracts were awarded to Boston College and the University of Virginia for the development of case studies for use as the focal point for group activities on Monday through Thursday. Each case was designed to demonstrate how a major project management decision was made or an action taken on a specific project. Furthermore, each case was to focus on one primary issue or decision related to topics discussed earlier that particular day. Cases would also provide a vehicle for codifying NASA's historical experiences, in order to transfer knowledge of lessons learned to new project personnel.[4]

Once the individual session had been established, it was necessary to select session leaders and panel participants. The commitment of NASA's top management to the program assured an outstanding group of trainers. NASA's first Agencywide training program was taking shape.

Program Implementation

The pilot NASA Project Management Shared Experience Program was delivered using three major segments. The first provided experiences directly related to the application of basic project management skills. Topics included NASA's approach to project management, project formulation, budgeting, procurement, contractor relationships and administration organization and execution, management of technical performance and management of experiments and software. The second segment provided a sequence of panel discussions on the role of the Program Office, industry's view of NASA, user/science community relationships, the impact of the Shuttle on project management and a four-hour Project Manager Workshop.

A third segment was devoted to group activities, including case studies and a task-oriented team development workshop. Case studies were designed to illustrate the effective or ineffective application of project management skills on past NASA projects. Participants worked in teams to analyze key project management issues, identify causes of project problems, and recommend actions. The cases contributed to the participants' abilities to understand the interdependence of business and technical decisions, and the need to understand the user and contractor environment in planning and executing projects.

The case studies provided NASA with a rich resource of documented past experiences.

The program was presented on May 2-7, 1976, in Reston, Virginia. A participant evaluation of the pilot program indicated that it was extremely successful, with an overall rating of 8.8 on a scale of 10. Written comments reflected the participants' positive opinions in terms of the significance of the topics and the effectiveness of the program's shared experience approach. The effectiveness of the shared experience approach in communicating concepts, procedures and techniques critical to project success was proven. And the NASA faculty seemed to value the program as much as the participants.

With the successful completion of the pilot, the LCSO had scored a management coup. Under their leadership, a major system training program had been designed, developed and delivered. The program employed large numbers of the NASA work force. Employees at all levels from assistant managers to NASA Associate Administrators were involved in some portion of the program.

This was the model that needed to be followed in the balance of the low cost effort.

The Project Management Shared Experience Program was presented twice a year from 1976 until 1980, when it was terminated because of a lack of a strong sponsor within NASA. On the job training (OJT) remained the NASA training process of preference. And there is much to be said for this method, especially if it is started early in a career and if it is accompanied by job rotation and mentoring. But at NASA, managers selected for billion dollar programs frequently had no project management experience or training and were forced to rely on OJT that came too late. The chaos that followed was predictable, for on the job training is not appropriate for the director of a program of national importance. In 1984 the

NASA Project Management
Shared Experience Program
Calendar of Activities

Time	Sunday	Monday	Tuesday	Wednesday	Thursday	Friday
7:00 AM 7:30 AM 8:00 AM		*Breakfast*	*Breakfast*	*Breakfast*	*Breakfast*	*Breakfast*
8:30 AM 9:00 AM 9:30 AM		Project Management In NASA	Procurement Process	Organization and Execution	Manned Systems Technical Performance	Project Manager Workshop
10:00 AM 10:30 AM 11:00 AM 11:30 AM 12:00 PM		Agency Budgeting	Contract Admin & Govt Monitoring / Cost Sched Perf & Mgt Reviews	Low Cost Activities	Unmanned Sys Technical Performance	
12:30 PM 1:00 PM		*Luncheon*	*Luncheon*	*Luncheon*	*Luncheon*	*Luncheon*
1:30 PM 2:00 PM 2:30 PM 3:00 PM	Registration	Project Formulation	Contractor Relationship	Management of Software	Management of Experiments	Conclusion
3:30 PM 4:00 PM 4:30 PM 5:00 PM		Case Study	Case Study	Case Study	Case Study	
5:30 PM 6:00 PM	*Free Time*	*Free Time*	*Free Time*	*Free Time*	*Free Time*	
6:30 PM 7:00 PM	Orientation	*Dinner*	*Dinner*	*Dinner*	*Dinner*	
7:30 PM 8:00 PM 8:30 PM 9:00 PM	*Dinner* Career Development	Program Manager Panel	Industry Speaker		Impact of Shuttle on Project Management	

PMSEP was reinstituted, but with only 50 participants a year and a limited curriculum offering, it was not able to meet the explosive training demands of the late 1980's. It also lacked the support of top management it had previously enjoyed, and was not able to attract committed instructors or participants.

The PMSEP had been designed and developed to meet the project management needs of NASA's work force of the future. It looked outward, toward the management of complex technologies and sophisticated missions while integrating lessons learned from the past. The PMSEP had effectively combined knowledge, technique and theory with an honest look at previous mistakes and successes—a format that was applauded by participants, faculty and the managers who had supported it.

Ironically, the PMSEP languished because it emphasized the development of human resources over tinkering with hardware. NASA would never allow a human crew to execute a spaceflight mission without extensive training, including hours and hours of mission simulations. NASA would not turn over an Apollo spacecraft or a Shuttle to less than a fully trained crew. And yet post-Apollo NASA management had no compunction against turning over multibillion-dollar programs to personnel who lacked the management knowledge, training, skills and experience needed to succeed. Post-Apollo NASA clearly saw in its future the repetition of Apollo-like successes, expertly engineered by technical geniuses. What it failed to see was that those successes would require management genius as well, that the technical experts would have to be trained and that resources would have to be spent to develop managerial expertise as well as spacecraft.

Footnotes

[1] Dale D. Myers, letter to Dr. Robert R. Gilruth, March 23, 1970.

[2] Robert R. Gilruth, letter to Dale D. Myers, March 25, 1970.

[3] Angelo Guastaferro, memorandum to Vince Johnson, Summer 1975.

[4] Dr. Michael Vaccaro et al., report of Project Management Shared Experiences Steering Group. Greenbelt, Md., Goddard Space Flight Center, December 1975.

Chapter 10

Uncovering the NASA Culture

From the very beginning of the Low Cost effort, work force motivation was thought to be essential to bring about a successful change. NASA had good experience with a motivation program known as Manned Flight Awareness.

This program had originated at the U.S. Army Ballistic Missile Agency at Huntsville, Alabama. In 1959 the decision was made to use a modified Redstone missile as the launch vehicle for the nation's first manned space program, Project Mercury. Contractor and government employees working on Redstone realized they would be responsible for a human payload. To ensure mission success and astronaut survival, the highest quality of workmanship during manufacturing, assembly and operations was necessary, and every effort would be made to obtain the best materials and people for the job. Much of this awareness was initially self-directed.

Out of this effort grew the Mercury Awareness program. This program featured a visual identification of all parts and documentation used for the Mercury mission. A special stamp or tag was placed on all components and parts to be used on the human-rated booster. These were to be the most perfect of the lot, and all identified items received individual inspection rather than lot sampling. The program proved very successful.

When NASA was formed, many of the Redstone Arsenal employees transferred to the Marshall Space Flight Center. They brought with them their experiences in quality, reliability and motivation and initiated a program of awareness for the relatively new Agency. The stated objective of the Manned Flight Awareness program was the assurance of mission success and astronaut safety, although program details would vary with the mission and goals of individual NASA Centers. The program used posters, literature, displays, speakers and films to both inform and inspire. Also, employees selected for special recognition were invited to attend a launch, the highest award in the program.

Since the program's inception, industry has participated at the prime, subcontractor and supplier levels. NASA has been pleased with the success of this program.

Building on these past successes and inspired by the informal NASA motto of "Can Do," the Low Cost Systems Office began to implement a program in 1974 aimed at persuading every NASA employee to "think low cost." But there were several obstacles to overcome. First, as an Agency of the Federal government,

NASA had to participate in an annual government-wide cost reduction program. Annual goals were set for all departments and agencies, and the goals were always met. Employees frequently referred to this cost reduction exercise as "the grand liars contest." This undisguised waste of effort left a very bad taste in the mouths of those forced to participate. Initially the LCSO was mistakenly identified with the government-wide program.

Next, and much more importantly, the project management work force had grown up with the expression, "They'll forgive you if you overrun, but they'll never forgive you if you fail." This saying was based on the NASA practice of instigating a "Failure Review Board" to investigate project mishaps. To many in the project management community, these boards were, at best, NASA's version of the Spanish Inquisition or, at worst, a military court martial where the tone of the proceedings was usually along the lines of "Bring in the next guilty bastard." Since projects represented the mainstream of NASA's business, it was important to ensure the project managers' full participation.

It was thought that a major cornerstone of any motivation program should be a significant cash award for low cost performance. A new approach was explored within the NASA Project Management community. Project managers were asked to consider the following: If they could share in a percentage of the "savings" (cost at completion below an agreed-to target), would that motivate them to work more cost effectively? To a manager, the answer was *NO!* Most thought this proposal would add one more level of complexity to an already overly complex job. Imagine, for example, two weeks before launch, the team is a million dollars under target and a potential problem is uncovered. Now the project manager must decide to conduct extra tests, wiping out the "savings" and the team's hard won share, or launching with an unacceptable degree of risk. The proposal was dropped.

To motivate the contractor community, a number of ideas were floated, including shared savings, the "E for Excellence" flags of the World War II era, and a trophy.

None of these happened, the difficulties associated with establishing appropriate criteria and judging the winner were never overcome. This was a most fortuitous outcome, as the industry trophy design selected by NASA was possibly the ugliest piece of ornamental art ever created. Imagine the conflict for the winning company. The good news is you've just won the NASA Low Cost Trophy; the bad news is that you are expected to display it.

Del Tischler and George Low with one of Johnny Hart's cartoons.

Photo courtesy of NASA

146

But some things worked. A popular motivational campaign involved a cartoon character called "Mighty" Low Cost, developed inhouse by NASA. Also in collaboration with Field Enterprises and Johnny Hart, one of America's favorite cartoon humorists and a friend of NASA, an equally popular "Quote Low" poster series was developed. The posters were continually purloined soon after they were put up. "Mighty" Low Cost appeared regularly in the *NASA Activities* magazine.

LOW COST CORNER

"MIGHTY" LOW COST SAYS

"USING PROVEN OFF THE SHELF HARDWARE, SAVES TIME AND MONEY!"

Redrawn courtesy of Maria Killingstad

However, from the beginning it was understood that outside professional help would be needed to design and implement an effective low cost motivational program. When it was finally decided to obtain help, Dave Berlew's name was first on the list.

Dr. David E. Berlew was then consulting with the General Motors Institute. He had received a Ph.D. in social psychology from Harvard University in 1960 and had served on the faculty of the MIT Sloan School of Management. He had been the Peace Corps in-country director for Turkey and Ethiopia while on leave from MIT. He was, in 1975, president of Development Research Associates, and his clients included Exxon, Honeywell, IBM, British Petroleum, the United Nations and the Island of Curaçao. He had the reputation of being a no-nonsense, straight shooter.

Low cost poster. *Courtesy of Norman Gerstein*

However, before he was brought on board he had to undergo a rather thorough vetting by the NASA Administrator's office. Even in the era of enlightened management, psychology and psychologists could be trusted just so far. Berlew passed.

Once Berlew had been okayed, he was quick to start an analysis of the low cost motivation problem. He visited two Field Centers, met with dozens of Headquarters personnel and attended a number of Agency workshops and meetings. He presented his written report on October 30, 1975, which began:

There is no "motivation problem" at NASA—not in the usual sense. Most NASA personnel want to do a good job and are committed to the Agency. Nor is there any shortage of available energy to get the work of the Agency done. However, there is more wasted energy than one would expect in an organization where members are so loyal. Energy is channeled away from productive work into a variety of areas, including: efforts to protect or cover oneself, to avoid getting into trouble, attempts to justify one's behavior, to oneself and to others, alternatively complaining about top management and asking for direction and support, trying to figure out the criteria by which people are evaluated.

Berlew indicated that wasted or non-goal-directed behavior is common in organizations which are in a state of flux or disequilibrium. NASA's reorientation to a cost conscious organization was, of course, a major cultural change. Berlew described the changes as:

- A change in orientation, from "accomplish the mission, damn the cost," to "accomplish the mission at the lowest possible cost,"

- A change in philosophy, from "do it the best way" to "make sure it works,"

- A change from an organization dominated by professional values to one where business values are equally important.

Berlew noted:

Cultural change touches upon the value of its members and affects basic organizational norms. This change causes confusion and stress and at NASA . . . members are told to think and behave in new ways, but get confusing, often conflicting messages about what those new ways are; they see inconsistencies between the policy statements and managerial actions of their leaders; they see differences between what they are told they should do, and what they and others are rewarded for doing; they realize that some of the old ways of doing things are no longer acceptable, but the alternatives are unclear or unattractive.

Berlew pointed out that cultural transitions are often accomplished by change in membership: the old guard is moved aside or leaves, and the new generation, representing the new culture, takes over. However, due to funding and Federal regulations, NASA would not have a new generation for at least another 10 years. The result was a work force that wanted to do a good job but may have been confused as to what the management expected of them. Or the system, which had yet to change, got in their way. This led to behavior driven by a lack of direction.

NASA management could do three things to facilitate the transition. First, do a better job in clarifying what the new mission is and communicate the rationale behind it. Second, identify and address the conflicts the new mission is causing for NASA employees. Third, ensure that the management practices then in force support the goals and values of the new culture, not the old. Managers must model their behavior to what they are asking others to do. Berlew felt that if these actions were effectively carried out, employees would understand what was expected of them.

During his visits to the various NASA facilities, Berlew identified several areas that needed attention to ease the transition and to establish a "new equilibrium."

A. There is a high degree of confusion and disagreement about the meaning and implications of "low cost" among senior NASA personnel.

1. Many people have defined recent changes at NASA as forcing a choice between low cost and quality and reliability. In their words:

"The major obstacle (to low cost) is no admission by top management that we will lower quality and reliability. We must have margin to take risks."

"The trouble is, George Low won't accept failure."

"Low won't take the risk of lower reliability."

"A 90% reliability goal leads to occasional failures. Headquarters' reaction is to stop the project, launch an investigation, and develop recommendations which are expensive to implement."

"NASA only talks one side of the equation—lower costs; it doesn't talk about increased risk. There is a communication gap with NASA top management on the risk question."

"A low cost award would be a reward for sloppy technical work."

A negative consequence of this perceived forced choice between technical performance and cost is that it "justifies" resistance to low cost efforts because low cost is against all that is "good," such as technical excellence, innovation and truth seeking.

2. There is confusion between form and substance with regard to low cost. Typical quotes are:

"If I were advising a new Project Manager, I would tell him it is more important to appear low cost than to be low cost."

"I don't see much beyond evangelism (about low cost) from where I sit."

"NASA is more interested in good performance against advertised price than low cost performance."

"Headquarters doesn't want to hear of cost growth; it is better to have a high estimate."

"Delays caused by Headquarters can cause loss of reserves through inflation. It ruins Headquarters' credibility regarding commitment to low cost."

"George Low makes speeches about low cost very easily and quickly. He doesn't know what he's talking about—he oversimplifies—as if we were manufacturing autos or TVs."

"NASA is dishonest; we've been forced into it by Congress."

"The same top managers who insist on low cost refuse to accept more risk in certain areas or a reduction in people reporting to them."

3. Low cost is viewed by some as conflicting with NASA's Research and Development (R&D) mission. For some NASA personnel, the emphasis on low cost raises the question about NASA's mission. They say that if part of NASA's mission is to push back the frontiers of science, then a low cost orientation will hinder mission accomplishment. If R&D is not part of NASA's mission, then it should be run like a business. In their words:

"I'd like to see clarification of what NASA is up to. Low cost can inhibit technological and scientific output and pushing out the frontiers of the state of the art."

"Low cost says: make it pay. But space doesn't pay. There is a basic conflict between NASA as an R&D organization and low cost. If we

really mean to put it on a paying basis, then drive down overhead and give bonuses to project managers and top people on cost performance."

4. People tend to associate low cost with a "paring" versus innovation. Many people see low cost solely as tighter cost controls, "creative" budgeting and avoiding cost growth. They say:

"A low estimate may really be high. We need a revolution in the way we do business. We do it at a very high cost."

"There is not much incentive to be creative about cost reduction. You can get by with paring down your budget."

"We haven't been in desperate enough straits to throw out old techniques."

"We are cut to one-third of our purchasing power, but we have kept our Centers and our people. A large percentage of our costs are institutional."

"We should use more parallel project teams—each trying to do it a different way. The problem is that it costs money to save money."

B. NASA's formal and informal reward systems do not consistently support, and often undermine, transition to a low cost culture. The nature and magnitude of this problem can best be communicated through direct quotes:

"Even in the last two years the people rewarded have not been in low cost programs."

"The punishment for failure is severe, but no project manager was ever pilloried because his project was successful but over cost."

"We haven't found ways to reward good cost performance."

"High cost but successful projects get criticized, but all is forgotten after launch."

"There is no reward for a technical person to get involved in a risk decision (e.g., whether or not a test can be eliminated). Where is the motivation for him? He is evaluated only on technical performance. So the project manager and his small staff become the cost drivers."

"Low cost puts stress on the project manager. He wants to succeed but has to do so with fewer resources. He is more likely to be called on the carpet for a mistake than applauded for success—especially if success is in low cost performance. Good low cost performance is mostly unnoticeable and therefore unnoticed."

"Being a low cost center doesn't insure that you will get work. It is just as apt to go to high cost centers because they sell or play the game better, or for political reasons."

"There is no relationship between advancement and recognition in the Agency, and low cost performance."

"The technical people force a project manager to take all the risks. They go on record as recommending more testing. That way they can't lose."

C. Blue ribbon committees and review boards are an expensive anachronism, inconsistent with a low cost orientation.

"There are too many reviews. I'm willing to stake my reputation on something but the review board comes in anyway. It takes a lot of everyone's time."

"Mistake investigations go too high. It is hard to delegate, but it should be done—to save money if for no other reason."

"Committees and review boards always take a conservative position: they recommend expensive tests and warn against taking risks."

"The review process is very expensive—in manhours, and in increased testing activity."

"Review boards and committees always recommend further expenditures to increase reliability. If each member makes recommendations, the project manager must dispense with each one because it is in the record."

Berlew next addressed specific actions to correct the disagreements and confusion he perceived over the low cost effort. He observed that even senior NASA officials were not clear about what Low wanted to accomplish. The frequent mention of Low's name indicated he was either the source of the confusion or the potential source of clarification.

In order to clarify the new mission objectives and plans and to reach a common understanding of the new culture, Berlew recommended that Low, Fletcher and their key subordinates get together for a day or two in a retreat atmosphere. The meeting should have a minimal structure to encourage a free exchange of ideas. Disagreements and conflicts should be encouraged as opportunities for learning. Participants at this meeting should conduct similar meetings with their staffs until a critical mass of NASA personnel understand and accept the low cost culture.

Berlew next recommended a careful analysis of NASA's structure to ensure that it supported the necessary changes to implement low cost policies. This would include an analysis of a group organizational structure and diagram including project and matrix organizations, the Headquarters and Center roles and mission, the integrative and coordinator mechanism, and the number and functions of the various levels of management. Also the analysis should include NASA promotion and compensation policies. This analysis should involve the best brains in the Agency and include enough representation to form the basis for the acceptance of change.

Berlew next identified "norms" in NASA which do not support or could even oppose formal new policies or procedures, norms being rules of behavior for members of the culture. If employees violate norms they are usually punished, often subtly and informally.

Two examples of these norms that did not support the new low cost culture were the reward system and the reliance on Blue Ribbon Committees and "old heroes." As to the reward system, Berlew found that neither formal nor informal rewards were used effectively to support the low cost effort. "In most cases, people do what they are rewarded for doing, not what they are told they should do," he noted. (This dilemma was to persist for two more decades, at least.) The reliance on Blue Ribbon Committees and "old heroes" or the "grand old men of NASA" might not be recognized as anachronisms in the new low cost culture but Berlew perceived them as conservative, expensive and a throwback to the old era. No matter how illustrious their past accomplishments, they were carriers of the old culture and very powerful players because of the respect the Agency accorded them. Berlew observed:

> As an example, consider the Project Manager Workshops, held under the auspices of the ad hoc Committee on Management of Space Program Costs of the Aeronautics and Space Engineering (perhaps all carriers of the old culture). Each Workshop was attended by several project managers, a few Headquarters personnel, a few outsiders and two or three "grand old men" of NASA as ad hoc committee members. I am not questioning the value of the sessions, because there was an exchange of ideas between NASA Centers about cost drivers and how to manage space projects, and this is an important contribution. At another level, however, it was a discussion of low-cost approaches (i.e., a "new culture" discussion) guided by heroes of the old culture.

Berlew observed the following at the sessions he attended:

- Business-related issues were occasionally brought up by project managers, but rarely explored or discussed by committee members.

154

- Project managers were more oriented toward gaining the approval and avoiding the disapproval of the "grand old men" than in solving problems.

- Senior members cross examine, and project managers defend, almost always with regard to technical problems and approaches.

- Project managers rarely said anything controversial or critical of NASA's low-cost philosophy or practices, although in private interviews they expressed considerable criticism and concern.

"This is not meant to be critical," noted Berlew. "The ad hoc committee members were simply reflecting the norms and values under which they operated and succeeded brilliantly in the NASA of the 1960's. One should not expect them to be at the cutting edge of a new, low cost culture."

By way of contrast, Berlew described the Director of Projects at one NASA Center as a "trendsetter," someone who is creating a new culture: "He has developed his project managers into well-rounded businessmen as well as technical project managers. He meets with them daily, as a group. He takes their word on technical matters, but works business problems with them. People are now interested in business!"

In conclusion, Berlew wrote, NASA should give high priority to:

- Specifying clear and credible yardsticks for measuring low cost performance

- Finding practical ways of rewarding low cost performance

- Giving greater visibility and influence to a new generation of "norm setters"[1]

The Berlew report was considered a very hot potato at the time and was held back for almost a month while the collective management of the LCSO steeled themselves for the confrontation they were sure would occur with George Low. When he was finally presented with the report, Low's reaction was radically different than expected. Low wrote a note to Gray to the effect that he agreed with almost everything Berlew had reported, and he wanted to move out quickly to correct problem areas. Unfortunately, Low had by that time decided to accept the presidency of Rensselaer Polytechnic Institute, a post he filled in the spring of 1976.

Consequently, few if any of Berlew's observations, recommendations or suggestions were acted upon. When Low left NASA, Berlew's report and the low cost effort rapidly faded away.

Footnotes

[1] David E. Berlew, Creating a Low Cost Culture Within NASA. Development Research Associates, Cambridge, Mass., October 30, 1975. Order for Supplies and Services No. W-13, 872.

Chapter 11

NASA Payload Policy: Facing Risk

One of the first studies conducted by NASA to explore low cost concepts was the "Payload Effects Analysis" completed in June 1971 by the Lockheed Missiles and Space Company. Del Tischler had been involved in this study and stated it was crucial in affirming his conviction that NASA could control hardware costs.[1]

The payload study was commissioned as part of an integrated economic analysis by three contractors of the new space transportation system. The specific objective of the Lockheed portion of the study was to develop concepts to reduce the costs of payloads developed for Shuttle launch.

The study found that, historically, payloads were designed within limited weight and volume constraints, with high-density packaging, heavy emphasis on low-risk hardware and extensive reliability and qualification testing. Payloads had a high level of documentation and configuration management associated with their development, and lengthy ground checkouts involving large numbers of personnel. Costs were driven by volume/weight limits, ground and flight test philosophies, repair or refurbishment approaches, risk philosophy, operating life, quantity and quality of parts, and the use of developed or qualified hardware.

The study identified three principal economic benefits for future Shuttle payloads: development of payloads using low cost design techniques, refurbishing payloads on orbit, and use of standard payload systems or standard spacecraft.

The study concluded that cost benefits would occur from new payloads designed to "low cost criteria" and that significant savings were possible with standardization of payload subsystems and the use of standard spacecraft. Some degree of savings could also result by implementing low cost practices for payloads launched on expendable launchers.

It can be argued that these were the studies that initially inspired Low to initiate the low cost effort. Payloads continued to be an important focal point for LCSO, culminating in the institution of a NASA policy recognizing that in the Shuttle era, all payloads should not be treated equally. The use of the Shuttle as NASA's primary payload launcher would give Agency management a previously unavailable opportunity to balance program cost and complexity with a factor of confidence in mission success. This logically would dictate "classes" of payloads ranging from those with a very high confidence level to those with more moderate levels. A direct cost correlation could be associated with the payload classes. The need for such a policy was first seriously debated during the GSFC Low Cost Workshop of 1975.

In November 1976, John Naugle, NASA Associate Administrator, issued a draft payload policy prepared by the Low Cost Systems Office. The policy would have the Associate Administrator assign each payload to one of four proposed categories of "desired confidence for mission success."[2] Each category contained general guidelines for program implementation. The categories would be assigned during the Flight Project Review.

The Flight Project Review process was also new. It too had grown out of a discussion at the Low Cost Workshop of the need for management to share the responsibility for cost reduction, an issue that had originally surfaced during Low's visit to GSFC in February 1973.

At Low's direction the policy was drafted and on June 15, 1976, John Naugle signed it. He began:

> At the Low Cost Workshop held last year with the Deputy Center Directors, the Deputy Administrator asked that a review be conducted in which the managers of "new start" flight projects would discuss with NASA management the basis upon which the project will be structured and managed. The purposes of this discussion are (a) to reach a common understanding of the low cost practices being incorporated into the project, (b) the level of risks involved, if any, (c) an agreement on how the project will be run, and (d) the basis on which the project manager will be evaluated during the course of the program.[3]

The review would cover major flight projects identified as budget line items and other selected projects. The review would center on mission objectives, the specific actions to reduce costs, the use of standard equipment, the management and procurement approach, and the cost/confidence/risk tradeoff options available. Other items impacting costs could also be covered. These include quantity of the hardware to be procured, schedule selection, spares philosophy, R&QA plan, contract features to reduce and limit costs and the major cost drivers, such as level of program definition, thoroughness of specifications and documentation, state-of-the-art improvements required, and degree of redundancy incorporated into the preliminary design. An agreement would be reached on how the project would be conducted and how the performance of the manager would be evaluated.

The Program Associate Administrators would be expected to conduct similar reviews for other projects they deemed necessary. Naugle asked that the results of these reviews be reported to him.

Each project would have a definite agreement between the Associate Administrator, the Program Assistant Administrator, Center management and the project manager as to specific implementation plans. Although not spelled out in the draft, this agreement reflected the desire of project managers to have some degree of shared responsibility if actions they had taken to reduce cost would result in failure. Low had promised this protection to the project management community. In reality, however, for a project lasting five years or more, the only managers left over from the original project would reside at the Center; most Headquarters managers would have been long gone, owing to the vagaries of Washington politics and individual ambitions.

Meanwhile, Naugle's November payload policy draft created a flurry of responses, most in general agreement on the need to classify payloads, but few in agreement on the way Naugle proposed to do it. Many of the responses were similar to the following, complete with the killer phrase "however."

"As stated in our earlier assessment, we feel that this process is an excellent approach to payload development. However. . . ." [4]

"We believe this policy will provide an important basis for discussions between the program officers and project personnel in the preparation and approval of project plans. However. . . ." [5]

"We strongly endorse the issuance of a NASA policy on payloads that will promote the economical utilization of the shuttle. The redirection from the 'success oriented' approach for space experiments to the more economical, but higher rIsk, 'design to cost' approach can be achieved only through the clear and visible support of NASA top management. However. . . ." [6]

"The policy statement is good. However. . . ." [7]

If anyone doubted the impact on cost of NASA's "no-failure" policy and the project managers' fear of Failure Review Boards, the response from James Kramer, the acting Associate Administrator for Aeronautics and Space Technology, should have set the record straight. In his reply to Naugle and repeated to Gray, he stated,

The mission characteristics of flight experiments listed . . . should include a line dealing with failure review requirements. For category I, the requirement could be a committee reporting formally to the Administrator; for category II, a program manager reporting formally to an associate administrator; for category III, a principal investigator reporting formally to a program manager; and for category IV, a principal investigator reporting informally to the program manager. The inclusion of this line provides NASA hardware personnel an assessment of the planned reaction of NASA management to failures and thus tends to reduce the perceived career impact of flight failures, thereby providing an incentive to assume greater risk, as required by their successful implementation of the policy. The brief treatment of failure closeout in the Develop, Build and Qualify section of the policy deals with test and pre-flight failures, and though appropriate, does not address planned NASA reaction to flight failures.[8]

After much debate, an Agency policy was released on June 20, 1977. The policy recognized the possibilities for reducing cost with the availability of the Space Transportation System; that past costs were driven by requirements to assure high technology; and that complex, irretrievable, schedule- and weight-critical spacecraft had to perform to full specification on their first and often only flight. Also, the project management approach developed by NASA could be characterized as schedule and performance intensive, with extensive development and qualification testing and rigid imposition of many detailed specifications, plus intensive monitoring, management and review of each contractor. This intensive management approach had been highly successful and was crucial to many of the Agency's past successes, but it was expensive to implement, and with the advent of the Shuttle, some projects could be treated with less management rigor than others. This adjustment had to be selective because the reduction of assurance for payloads depended on the transportation: irretrievable payloads would continue to demand a high level of assurance activities, but attached payloads, only a minimum level.

The established payload categories, arranged in order of descending requirements for assurance, were:

- Upper stage payloads

- Directly deployed payloads

- Attached experiments; schedule constrained or critical

- Attached experiments; space available

It was the intention of the new policy to provide substantial latitude to programs and projects in the planning of assurance aspects. The implementation process stipulated that each instrument should be designated in the Project Plan as either Class A or Class B in accordance with the following criteria:

Class A, Maximum Confidence—Instruments, and their required spacecraft systems, which justify a maximum confidence approach in accordance with one or more of the following criteria:

- To assure achievement of objectives of high national significance.

- To avoid prohibitively expensive consequences of failure.

- To preclude hazardous failure modes.

Class B, Nominal Confidence—Instruments, and their dedicated spacecraft systems, which are cost constrained to the extent that acceptance of failure risk is considered acceptable when it is cost effective.

During the Flight Project Review, the class designation will be established and serve as the basis for delegation of payload failure review responsibility. Failure reviews of all Class B equipment on attached payloads will be instituted by and report to the Center Project Manager; for Class B equipment on free flyers, the Center Director; for Class A equipment, the responsible Program Associate Administrator.

The Flight Project Review covers the design approach and activities to develop, integrate and operate the payload elements. Special emphasis would be given to the assurance aspects of the program.[9]

In January 1977, as a follow-on to a Low Cost Shuttle Payload Workshop held the previous fall at the Marshall Space Flight Center in Huntsville, Alabama, Ed Gray wrote to 18 aerospace corporations requesting their thoughts on "new approaches" that would reduce payload costs. The companies that replied provided the LCSO with some valuable suggestions. A summary of the most useful responses follows.

- Exploit Shuttle capability to retrieve and repair payloads in the design, test and acceptance of payloads.

- Move toward standardization of as many elements of smaller payloads as possible.

- Standardize more on components, spacecraft, etc., and reduce testing after development and strive less for design optimization when it drives cost without increasing return.

- Group payloads into common packages to exploit similarities.

- Make contractor responsible for specification definition and compliance.

- Specify interfaces in detail and use mini-specifications.

- Place responsibility of the payload side of the interface solely on the payload developer.

- Reduce complexity of interfaces between payloads, Spacelab and Shuttle.

- Standardize procedures for requirements, acceptance, receiving, assembly and test of payloads.

- Build equipment/spacecraft in production lots.

- Plan more flights with less sophisticated payloads.

- Reduce mission success criteria for returnable payloads.

- Develop a standard set of payload services.

- Classify payloads into four classes so that cost tradeoffs can be rigorously defined. Class I is "mission critical," Class IV is a "convenience payload," and Classes II and III are in between.

- Concentrate on achievement of science and applications objectives, not equipment development, since we will have frequent flight opportunities and payload recovery capability.

- Centralize responsibility for data management.

- Utilize payload developer through integration, flight and post flight cycle.[10]

Ultimately, a number of these suggestions were implemented in some manner. For instance, NASA did initiate a formal payload classification policy designed to save costs, but as those who attempted to follow the low cost approach soon found out, all payloads flown on the Shuttle became, for all practical purposes, "mission critical" once they arrived at the Cape. The "no failure" culture quickly canceled out the low cost policy when the payload approached launch.

Footnotes

[1] NASA defines payloads as "an aggregate of instruments and software for performance of specific scientific or applications investigations or for commercial production. A specific complement of instruments, space equipment, and support hardware carried into space to accomplish a mission or discrete activity in space."

[2] John E. Naugle, Memorandum: "Policy on NASA Payloads," June 20, 1977.

[3] John E. Naugle, Memorandum: "Flight Project Review," June 15, 1976.

[4] Bradford Johnston, Associate Administrator for Applications, Memorandum dated January 13, 1977.

[5] James J. Kramer, Memorandum: "Proposed Policy on NASA Payloads Categorization." Washington, D.C., January 17, 1977.

[6] Donald P. Hearth, Director, Langley Research Center, Memorandum dated December 2, 1976.

[7] Hans Mark, letter to John Naugle dated November 22, 1976.

[8] James J. Kramer, Memorandum: "Proposed Policy on NASA Payloads Categorization." Washington, D.C., January 17, 1977.

[9] John E. Naugle, "Policy on NASA Payloads," June 20, 1977.

[10] Edward Z. Gray, Memorandum: "Industry Views on Opportunities for NASA to Reduce Program Costs," March 17, 1977.

Chapter 12

Evaluation: Was It Worth the Effort?

In the fall of 1977, the future of the Low Cost Systems Office was as uncertain as ever, even though by this time some in NASA had actually bought into the idea of reducing costs. It was, after all, an excellent way to demonstrate to the Office of Management and Budget and Congress that NASA was a good steward of its allotted resources. But the office had few supporters in the NASA hierarchy, and after the departure of George Low, its opponents felt unleashed. NASA Center Directors were especially concerned. Developing standard equipment might be good for NASA, but there was nothing in it for the managing Center. This attitude was best demonstrated by a Center Director and personal friend of Low's who, in response to a query on what support he planned to give this important new effort, replied, "None! There's nothing in it for me." He allowed that if Low gave his Center $10 million, he would do something significant, but without that level of funding, he would do nothing.

What the Center Directors saw was not the need for change, but additional work that, for the most part, brought no new resources to their Centers. If the standard equipment they managed was late, it could adversely impact a sister Center's project. As we have seen, reducing cost and saving money were never important in the NASA culture, and with the departure of Low, they dropped even further on the Center Directors' priority list.

The continuance of the low cost effort was also being questioned by the new NASA leadership appointed by the Carter Administration. In an effort to determine the efficiency of the Low Cost Systems Office and its proper reporting responsibility, Harry Sonnemann was tasked by the new NASA Administrator to conduct a detailed review. Sonnemann, a former special assistant to the assistant secretary of the Navy for research and development accompanied Robert Frosch, on his journey from the Navy Department to the NASA administrator's job in 1977. Born in Munich, Germany, and educated at the Polytechnic Institute of Brooklyn, Sonnemann was the assistant director at Columbia University's Hudson Laboratories before joining the government. He held a number of significant positions over the next 13 years with the Advanced Research Projects Agency, the Nuclear Test Detection Office and the Navy Department. The findings he reported on December 21, 1977, were long on words and short on hard recommendations. His report addressed four principal issues:

- Had the standardization effort been effective in reducing space system costs?

- Should the entire effort be continued?

- What major changes were indicated, if any?

- What were the options for the organizational home of the Headquarters component?

To answer these questions, Sonnemann interviewed more than 50 NASA Headquarters and Field Center leaders, plus representatives of seven aerospace corporations, and pored over much of the office's documentation. He began his report with a review of the office's brief history and accomplishments. Over the five-year period from FY1974 to FY1979, $33 million had been allocated to the low cost effort with 22 pieces of equipment declared NASA standards. The number of civil service employees devoted to the low cost effort varied according to each Center's involvement in developing standard equipment and supporting the panel and practices activities. Goddard, Marshall and JPL devoted the most personnel for the period FY75 to FY78 to the low cost effort.

Resources Summary—Manpower Estimates by Field Center

	FY 75	FY 76	FY 77	FY 78
Civil Service				
Ames Research Center (ARC)	1	1	–	–
Goddard Space Flight Center (GSFC)	19	17	30	28
Jet Propulsion Laboratory (JPL)	12	8	20	24
Johnson Space Center (JSC)	–	–	1	–
Kennedy Space Center (KSC)	3	3	–	–
Langley Research Center (LaRC)	–	–	–	3
Lewis Research Center (LeRC)	7	10	4	4
Marshall Space Flight Center (MSFC)	10	10	18	18
	52	49	73	77

Sonnemann's report included standard equipment market projections based on the latest approved NASA payload model. For instance, a projection of the potential standard tape recorder purchases through CY1980 was compiled. A total of 86 recorders were identified, a very large potential future market.

The standard transponder was used to illustrate typical savings realized by adopting a standard. This was appropriate in that the NASA standard transponder was to become one of the most successful items of standard equipment development, versions of which are still in use today, 20 years after its initial development. The chart showed a potential $15 million savings for a purchase of just 20 standard transponders, an outstanding return on investment when the average cost of a single unit fell from over $1 million each to approximately $200,000 each. This was an incredible bargain for NASA and all future users.

Sonnemann then used a cost/benefit assessment done in August 1976 to show the scope of the potential cost savings of the total effort. In private industry, potential savings such as that shown in the following chart would have been more than enough justification to continue the program, but in government, saving a few million dollars may not always justify the effort. It was not an important goal of the organization.

Total Low Cost Systems Cost Benefits

Practice	Five Year Cumulative $ Mil
Available Equipment Use	25.0
Reduced Documentation	12.5
Less Specifications	12.5
Decreased Testing	25.0
Less Spare Parts	12.5
Improved Flight Operations	3.5
Cost and Pricing Data Bank	2.5
Cost Simulation Modeling	.4
Subtotal	93.9
Standard Components	165.0
Total	258.9

Assumptions: R&D for unmanned Flight Projects $500M/Year
5 Year Benefit Accrual Before Block Change

Sonnemann now got to the crux of his study, a listing of 43 findings. Although his final recommendation was to continue the LCSO effort, the sum and substance of the report were not a ringing endorsement of that position.

He structured his report along the lines of the LCSO organization. A sample of his findings and conclusions follows:

Standardization Program

- The standardization effort will have measurable cost benefits, and identifying components, subsystems and systems with multiple mission applications will also minimize duplication.

- The low cost program is responsive to Congressional concerns about the high cost of space activities.

- The "first user as the developer" concept appears to be the only one which fully assures an interested technical organization committed to the completion of the standard development, and funding which can be defended against budget cuts.

167

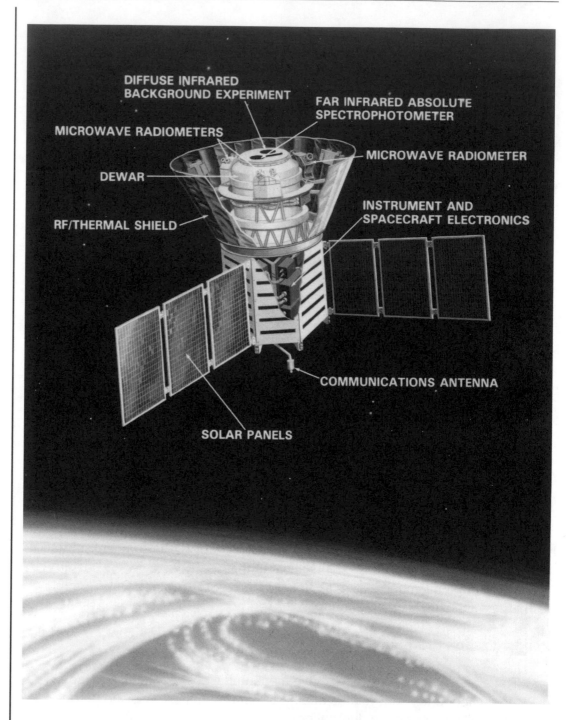

The NASA Cosmic Background Explorer (COBE) was launched on a Delta rocket in 1989 to study the origins of our universe. It was a most successful mission. The COBE spacecraft included two pieces of standard equipment developed by the Low Cost Systems Office in the mid-70's.

Photo courtesy of NASA

- To maximize opportunities to compete, all standards should be specified in a form, fit and function format with the interfaces defined. A number of the present standards need to be revised to remove either Center design biases and/or unique design parameters which restrict the field of potential offerors. Since market potential and rate of utilization of NASA standards tend to be low, development of standards must be approached carefully. The objective to minimize the cost of space equipment may frequently be better served by reutilization or adaption of flight hardware and/or certifying such hardware as a "pseudo" standard.

Business and Program Practices

- Industry indicated that DoD and commercial contracts place a far greater emphasis on on-orbit performance than NASA does. This encourages a higher equipment legacy and much closer industrial management scrutiny.

- A review of NASA and industry practices indicates a significant effort to reutilize and adapt previously flown space hardware. There is a significant equipment legacy from previous space programs within the framework of the standards program, and industry should be provided the opportunity to offer competitive items.

- All the studies reviewed indicate that there are opportunities to reduce costs by increasing the commonality of contract, reporting, testing, specifications, drawings, and other requirements. The estimated maximum savings are 10 percent of total program costs.

- Spacecraft costs correlate with the spacecraft's complexity.

- Modularity provides the flexibility to accommodate mission requirements and, therefore, the greatest cost savings due to ease of adaption of various payloads. It also provides a good potential for Shuttle in-orbit servicing and the greatest use of standardized equipment and industrial participation.

Conclusions

- The standardization program has been moderately successful and should be continued.

- The Agencywide nature of the standards and program practices efforts suggests that they should be located in an office which has Agencywide responsibilities. The office of the Chief Engineer appears to be the only management element under the new organization structure that has Agencywide responsibilities for technical execution of all NASA programs. This office can assure that development efforts and mission operations are being planned and conducted on a sound engineering basis with proper programmatic controls.[1]

Following Sonnemann's report, the LCSO had one last opportunity to make its case, a rebuttal to the Administrator. Its presentation mirrored Sonnemann's: a chronological history of the office, an outline of its charter and a review of the standard equipment process. Typical cost savings were also presented, along with the status of the standards then under development. A few problem areas were noted, mostly touching on the lack of technical and management expertise at both NASA Field Centers and among NASA's contractors. The presentation acknowledged the Field Centers' concerns regarding the lead Center role, the process for declaring a standard and the acquisition of standard hardware following initial development.[2]

Then in a bold if futile move, the LCSO presented a number of far-reaching recommendations.

The first would have the LCSO redesigned as the Standard Systems office, reporting to the NASA Deputy Administrator. As with Tischler's Task Force, having the office report directly to the Deputy Administrator would raise the office in stature and influence, which the LCSO staff felt was necessary to resolve intra-agency management issues.

The second recommendation would consolidate all standard equipment development, acquisition and logistics at the Jet Propulsion Laboratory. The consolidation would solve the lead Center and logistics problems, but it would cost more because JPL support charges were higher than those incurred at other NASA Centers. If top management wanted an aggressive, effective and responsive standards program, consolidating it at JPL would be a reasonable solution.

Recommendation three would establish a separate budget line item for the development of standard equipment independent of the first user—not an unreasonable request for a dynamic program.

The fourth recommendation would establish procedures for acquisition of standard equipment in consolidated procurements. This would solve the block-buy dilemma, achieve cost advantages and increase the potential for future success of the effort. The potential to do this definitely existed.

The fifth and final recommendation would have the Administrator reaffirm the goals and objectives of the program, including its new guidelines and directions. The LCSO hoped this would inspire top management to support the low cost vision as embodied in the LCSO program.

Unfortunately, none of these recommendations was accepted. The LCSO eventually was transferred to the NASA Chief Engineer, who was openly antagonistic to its leadership, its people and its goals and objectives. As Anthony Downs wrote in *Inside Bureaucracy*, bureaucracies will always work in their own best interest. The best interest of the NASA bureaucracy at that time was not well served by the Low Cost Systems Office. Although still in existence, its days were numbered; 394 to be exact.

Later evaluations of the accomplishments and effectiveness of the Low Cost Systems Office appear in widely divergent places.

The following evaluation was completed in the summer of 1985. It was conducted for John Sheahan, Director of Business Management of the newly formed Space Station program office, in the hopes that findings from the low cost systems efforts could be applied to the space station development. Thomas La Croix, a budget analyst, working for Sheahan, conducted the review. Portions of his report are as follows:

The central fact about the Low Cost Systems Office is that it primarily focused on the development of standard components for unmanned free-flyer spacecraft. Its impact was intended to be felt over time with a number of missions where standard off-the-shelf technologies could be introduced into spacecraft subsystem designs. These technologies could be provided as government-furnished equipment (GFE) to the prime spacecraft contractor. The contractor would also be induced with a variety of other procurement devices, especially by fee structure, to keep costs at a minimum.

The literature available suggests that standard technologies were expected to have a useful application within the industry of up to five years. The standard 10-to-the-8th tape recorder and the standard S-band transponder have been around a lot longer than that. They have had a long and glorious history of success in a number of NASA spacecraft. In all, when one overlooks the in-flight failure rate of missions closely identified with the LCSO, there was a clear trend of greater cost control, from initial estimate to final cost, for missions that were developed in the middle 1970s—that is, in the period in which the LCSO was most active.

However, the procurement practices that characterized the LCSO (such as a fixed-price incentive contract) still have much to suggest about how to keep costs down. An imaginative approach to the fee structure, particularly where both negative and positive incentives were employed, can be expected to work in NASA's favor.

Employing standard or off-the-shelf hardware from any source, LCSO or not, is inherently desirable as a potential cost saving device wherever a timely decision to employ it is made.

Practices that backfired on LCSO relate especially to the amount of test hardware and to test practices themselves. The protoflight approach may have evolved out of the low cost practices. Where extreme efforts were employed to reduce test hardware in using a protoflight approach (SEASAT and the Infra-red Astronomy Satellite, IRAS, for example), the result was disastrous either in terms of cost (IRAS) or of inflight performance (SEASAT). Where the protoflight approach was successfully employed, there was no great hesitation to employ adequate test hardware and procedures. Thermal, structural and mass models, breadboards, brassboards and full-up form, fit and function engineering models all were employed as needed.

The sole distinguishing fact was the absence of a full-up form, fit and function systems or subsystems prototype that was to be tested to destruction.[3]

Another point of view on standard equipment was provided by Albert D. Wheelon, the former Chairman of the Board of Hughes Aircraft Corporation and intimately familiar with spacecraft development. In an article "Space Policy Reconsidered," Wheelon stated:

There is a tendency to standardize space systems and subsystems. The argument for doing so is to spread the non-recurring cost of development over a large base and thereby avoid frequent investments in new designs. But one also pays a price for constraining different missions to use the same system. The alternative is to develop a very expensive system that suits all needs.

This important lesson was learned by the Air Force in its Inertial Upper Stage (IUS) Program, which was to serve as a universal payload carrier from Shuttle altitude to higher orbits. It began as a simple two-stage, solid propellant vehicle that would cost a few million dollars. When the requirements of all the missions IUS was to serve were imposed on the design, the recurring cost rose past $50 million. The IUS became the largest common denominator of all user requirements. The point to note is that the cost of adapting to many missions can easily overwhelm the savings of a common design.

The NASA standard module program of 10 years ago withered because it tried to freeze technology. It paid an enormous price for generalized solutions.[4]

And yet another view is available from the SEASAT Failure Review Board investigation. Completed in 1979, excerpts from the Board's report point out the benefits and pitfalls of standard hardware:

The advantages of using standard, well-proven equipment in terms of both cost and mission success are well recognized. But the experience of SEASAT illustrates the risks that are associated with the use of equipment that is classified as "standard" or "flight-proven." The uncritical acceptance of such classifications by the SEASAT engineering staff submerged

172

important differences in both design and application from previously used equipment. It is therefore important that thorough planning be conducted at the start of a project to fully evaluate the heritage of previously used equipment.

With the stimulus of the design-to-cost ceiling, and management emphasis on the maximum use of existing subsystem hardware, the Jet Propulsion Laboratory (JPL) Definition Phase Group proposed the idea of building a spacecraft system comprising two major elements: a sensor module designed specifically for SEASAT, and a spacecraft bus based on an existing, flight-proven bus developed for other Air Force or NASA programs.

Consistent with the concept of the "standard Agena bus" was the policy decision to minimize testing and documentation, to qualify components by similarity wherever possible and to minimize the penetration into the Agena bus by the government. As a result, a test was waived without proper approval, important component failures were not reported to project management, compliance with specifications was weak and flight controllers were inadequately prepared for their task. Significantly, the SEASAT slip ring assembly had no applicable flight history at the time of its launch and, in its application to the spacecraft, was a new device.

. . . It is the main lesson of SEASAT that an uncritical acceptance of such classifications as "standard" can submerge important differences from previously used equipment in both design and in application. It is important, therefore, to conduct thorough planning at the start of a project to fully evaluate the heritage of such equipment; identify parts that are standard and those that are not; and establish project plans and procedures that enable the system to be penetrated selectively.[5]

A cost benefit assessment of the LCSO standard equipment and practices program was conducted in 1976 by Anthony Diamond of the LCSO. The impartiality of the analysis can be questioned along with some of the assumptions, but Diamond was known as a conservative cost estimator, not easily swayed by management's desired outcomes. A portion of his report follows.

This report presents the results available to date of a cost benefit assessment being performed by the Planning Research Corporation (PRC) and the NASA Low Cost Systems Office (LCSO). The purpose of the assessment is to determine the dollar savings which accrue when the "low cost systems approach" is used in place of the "business as usual" approach in the conduct of NASA programs.

Not Built as Planned

A rupture in the fuel line of the Mars Observer resulted in a costly mission failure. Dr. Coffey, quoted in the Jan. 6, 1994, New York Times, said the Observer was patterned after Earth-orbiting weather satellites to save money. However, the Mars spacecraft was changed so much and got so complex that the failure was bound to occur.

Considering only the equipment standardization announced to date, the savings for a five-year period is in the range of $84 million to $165 million depending on the degree of conservatism employed in the cost benefit methodology. The savings will increase as further standardization is implemented. In the program and business practices area, the savings could be as high as $140 million for the five-year period.

For the particular components considered in this assessment, the results for the conservative approach indicate that use of the standard rather than the business as usual approach can reduce component total Development and Production Costs 18 to 38%. Obviously, the total amount of cost benefit is sensitive to the mission model which determines the number of units required. However, it can be stated that in general, for the components considered in this assessment, that the amount of cost benefit increases significantly as the number of units produced increases.

Additionally, it is anticipated that implementation of the multimission spacecraft program, as well as standardization in the other spacecraft system areas and in the payload instrumentation area . . . will add appreciably to the projected cost benefits.

The Low Cost Systems Office sponsored the design and implementation of the first Project Management Shared Experience Program held May 2-7, 1976. Benefits of the program include improved knowledge of approaches to the handling of cost and schedule drivers through effective planning and execution. It is anticipated that by transferring the Agency's "lessons learned" the program will lead to long-run savings of significant (but non-quantifiable) proportions to the Agency.

The total savings estimated for all of the Program and Business practices amounts to about $18.8 million per year. This represents a cumulative cost benefit for a five-year period of $94 million. This is a very conservative estimate because we have recognized the difficulty involved in effectively implementing these practices throughout the Agency. However, if this difficulty is surmounted through effective planning and execution, it is estimated that the five-year cumulative saving will be increased by as much as 50% to a realistically achievable savings of $140 million.[6]

From a purely cost savings standpoint the program had a good potential. Most of the assumptions, however, especially the traffic models on which the estimated savings were based, were very optimistic and were in actuality never met.

174

Footnotes

[1] Harry Sonnemann, "Low Cost Systems Activity Review," December 21, 1977.

[2] "Background to the Formation of the Low Cost Systems Office," a presentation made circa January 1978.

[3] T. LaCroix, Memorandum for the record, "Low Cost Systems Office File Review," July 23, 1985.

[4] Albert D. Wheelon, "Toward a New Space Policy" in Space Policy Reconsidered, ed. Radford Byerly Jr., Westview Special Studies in Science, Technology and Public Policy, 1989.

[5] NASA Investigation Board, SEASAT Failure Review Board Report. NASA, Washington, D.C., December 21, 1978. Reprinted in Readings in Systems Engineering, ed. Francis T. Hoban and William M. Lawbaugh, NASA Scientific and Technical Information Division: Washington, D.C., 1993. SP-6102.

[6] Anthony Diamond, "NASA Low Cost Systems Office Benefit Analysis of Standard Equipment," March 21, 1977.

Chapter 13

Culture and Bureaucracy Clash with Low Cost

The low cost effort was terminated for three major reasons: the resistance of the NASA culture, the debilitating effects of government bureaucracy, especially where cost cutting was concerned, and the lack of support from George Low.

The NASA Culture

The first obstacle, never to be overcome, was the NASA culture shaped and cemented by the Apollo Program.

When Low started the low cost effort, he may not have realized the full extent of the change he was seeking. He attempted to alter an organization accustomed to having more than adequate resources, being unique rather than "standard," and using management techniques that did not stress cost effectiveness, efficiency or centralization. In the early 1970's, the notion of organizational culture was almost unknown in NASA; Low knew little if anything of culture in a management or organizational sense until he read the Berlew report.

In their book, *Corporate Culture and Performance*, Kotter and Heskett remind us that "when the cultures are our own, they often go unnoticed—until we try to implement a new strategy or program which is incompatible with their central norms and values. Then we observe firsthand the power of culture."[1] This accurately describes what happened when the NASA culture met the low cost effort.

Culture is built upon an organization's way of doing business as well as its interaction and adaptations to internal and external environments that are perceived by its members to be successful. Members choose what events to remember based on factors they perceive to have contributed to success. These institutional practices and behaviors are reinforced if they are used repeatedly and if they seem to be related to other successes in the organization.

Memories of other ways of doing business are often not recorded and are lost. Also, competing interpretations of reasons for success are often not examined or evaluated, and special environmental conditions surrounding success are not fully considered. Over time, these perceived successful behaviors, organizational structures, power relationships, technical methods and other practices become entrenched as "the way" of doing business. To think otherwise or to consider other options is to adopt a mindset known as "not invented here" (NIH). Like any other organization, NASA's employees created their culture, and other successful methods or practices were simply NIH.

The practices, behaviors, assumptions and structures established at NASA during the Mercury, Gemini and Apollo years essentially created a culture that succeeded at, and therefore valued, human space flight. Since these programs existed during a time of nearly unlimited resources, widespread public support and technological advancement, the emerging NASA culture reflected its environment. Cost was not an issue, each system was considered one-of-a-kind and unique in every sense of the word, and management practices existed to facilitate progress, not as a means of control. Even the subordinate culture in NASA that supported "unmanned" programs regarded mission success more highly than any other factor.

The Low Cost Systems Office represented a radically different approach. Low knew that NASA's world was changing and that innovative approaches would be needed to ensure the continuance of the nation's civil space program. But for true cultural change to occur, as many organizational and management theorists contend, the equilibrium of an organization must be upset. Severe financial problems, like those recently experienced by the automotive industry, or a loss of important management leaders or organizational founders could be sufficient to force an organization's members to examine, question and change established practices.

In retrospect, there was no major disturbance in the NASA environment during the Low Cost System era. Many members of the organization who had created the dominant NASA culture were still in control, and they expected a grateful nation to give them another space spectacular to manage. Even with Low's leadership, the NASA culture could not comprehend the need to adopt new ways of doing business or to experiment with new methodologies or management practices. The organization was stuck in habitual ways of doing business that had previously brought stunning results. In less than a decade, NASA had become a common bureaucracy, entrenched by a high institutionalized cost, and intent on performance above all else.

The study of institutions tells us that most groups of people over time develop set ways of interacting, along with predictable patterns of behavior and similar mental schemes about the world. In NASA's case, the same group of people from the 1960's were still in charge in the 1970's, and there had not been a healthy infusion of new talent or fresh perspectives into the Agency. In essence, the mindset of those in NASA was "stuck" in old practices that had worked before, a mindset that sees innovation as strange and irrelevant.

The difficulty of selling low cost in such an environment was predictable. The powerful NASA culture—combined with bureaucratic inertia and the perceived lack of any major disturbance to provoke organizational learning—cemented "the NASA way" of doing things. In the presence of such a status quo, the low cost systems effort could hardly prevail.

For a more experiential view of the difficulties of implementing cultural change in NASA, it is helpful to look at the efforts of

Dr. Humboldt C. Mandell Jr., a long-time NASA employee of the Johnson Space Center in Houston, Texas. "Culture" Mandell, as he is sometimes referred to by his colleagues, is well known in NASA for his pioneering work on the effects of culture on program costs and his untiring efforts to promote positive changes to reduce cost.[2]

Mandell also observed that following the Apollo Program, the general population of NASA had no sense of any need to change. Why change? After all, they had just completed the greatest engineering triumph of mankind. The world had applauded. All were at the apex of their professions. Promises of even more exciting missions lay ahead, notably a human expedition to Mars. So why on Earth would anyone want to change a perfect system?

Ignored or denied in this perception was the rapidly declining budget. Most anticipated that the budget would again be increased when the need arose. Only a few forward thinkers saw that free-flowing money would not continue forever, and that if major programs were to succeed, lower costs were imperative.

Mandell worked with George Low for many years and thought Low brought unprecedented clarity and understanding to the NASA cost dilemma. Low's engineering background, however, gave him little appreciation for the overwhelming power of the prevailing culture. Although the presence of cultural influences on organizations has been recognized for more than a century, NASA managers, almost all of them engineers, universally treated organizational change as a simple problem. To change an organization, change the organization chart by moving managers around. These engineers, and hence NASA management, had no exposure to change based on cultural approaches. Mandell knew many would ridicule the theory that human feelings are data to be reckoned with in making organizational decisions.

And so George Low, despite his enormous intellect and his passionate desire to initiate change, was also a victim of the prevailing paradigm of NASA management. He may have been blind to the hidden forces at work, but he did recognize their results.

The NASA organization shielded most of its management from any centrally led change process. Each Center was a mini-NASA unto itself. Center Directors have often been compared to feudal barons, with near-absolute control over their Centers, and usually strongly supported by their Congressional delegations. NASA Headquarters traditionally moves very carefully when making changes that affect a Center. Additionally, despite Low's urgings, the Center Directors did not perceive a need to change. If George Low was saying that NASA needed to find new ways to save money, wasn't that a signal that the Centers needed to compete harder for their share of the shrinking budget? The culture created the image of greedy siblings fighting over the family fortune, rather than cooperating to achieve cost-saving organizational changes.

The expectation of the time was that the original "von Braun vision" was going to be realized, that Apollo would soon be followed by an extensive lunar exploration program, a space station and a visit to Mars. Since the Apollo management style had been highly successful, why change it for these new ventures?

Many perceived that the nation and the world had only begun to explore space, and that grand, visionary programs would continue.

The Space Shuttle Program

Mandell recalls that early in the evolution of the Space Shuttle, particularly when budgets did not emerge as expected, there was a frantic search by some in the program office to find ways to reduce costs without sacrificing program content. Within six months, dozens of organizations with reputations for low cost development, along with well-known NASA contractors, were visited and interviewed.

The findings were tough medicine; to reduce cost, NASA would have to completely change the way it did business with the private sector. NASA would have to forfeit any active participation in the development process, define what it wanted and leave the task of producing the hardware to the competitive private sector. NASA would have to use the contractors' reporting systems to avoid costly new paperwork. Above all, NASA would have to change its procurement systems. Under contract practices that still prevail today, NASA paid all developers a minimal fixed-fee based on total expenditures, with more fees awarded if milestones were met. Amazingly, the successful operation of the actual hardware was not a significant factor in the award process. This reward system enabled the contractor to make more money without necessarily producing the specified spaceflight systems and hardware.

When confronted with these facts, Shuttle program managers did nothing. They knew that they would be evaluated not on how much their program cost, but on whether the Shuttle flew successfully. They saw little to be gained from making changes. After all, cost savings would mean having to lay people off from their jobs, which would negatively affect morale and could result in some loss of performance and an attendant increase in risk. The management team was not inclined to increase risk just to save money. A safe flight was everyone's top goal and no cost was too great to assure that goal.

Interestingly, experience at NASA taught that "change" actually increased costs. A program manager would change a spacecraft to make it safer, or to achieve better performance, adding costs to the program. Contractors received monetary rewards for introducing changes; their fees were based on how much money the government spent. There were (and still are) few methods available to the government to resist this prodigal culture.

Consider the plight of any NASA program manager, who is presented with a change that the contractor believes is "mandatory" to assure the safe performance of a spacecraft. The program manager looks to an inhouse engineering expert for advice. Unencumbered by any responsibility for the program budget, that engineer would almost always advise the program manager to approve the change. Faced with the contractor and the NASA expert advising in favor of the change, what program manager would ever say: "We can't afford it"? Cost was not included in the decision-making process.

When presented with the fact that many of the changes to the Space Shuttle were increasing the cost of its eventual operation, the program manager said on several occasions, "Well, it [the operations phase of the program] won't happen on my watch." In other words, he knew that he would be measured only by the success of the early flights and by returning the crews safely, not by how much money was spent. And under the system existing at the time, he was correct.

In summary, Mandell and others believe that cultures change only when forced to change, and only to the extent dictated by the prevailing forces. In the absence of a unifying vision, the culture becomes vulnerable to whatever forces can shape its destiny. In any public institution, without a unified vision, the prevailing politics dictate the directions of change. Thus, Low and the Low Cost Systems Office didn't stand a chance.

Government Bureaucracy

The second major reason the low cost effort failed was the inherent nature of government bureaucracy. More than likely, bureaucracies have been with us since the earliest of times. In the early 1900s Max Weber, a German sociologist, used the term to describe a particular hierarchical structured organization in which authority is gained by legal and traditional means. Weber also identifies the importance of personal leadership. In the intervening years, the term has come to have a somewhat negative aspect. The *American Heritage Dictionary*, *Second Edition*, defines bureaucracy as "an administrative system in which the need to follow complex procedures impedes effective action." Although bureaucracies may have a negative connotation today, they are the management forms for many of the most powerful, important and influential institutions in this nation and the world.

During the halcyon days of Apollo, few in NASA would believe they were a part of a huge government bureaucracy. NASA's leaders and managers had worked hard at keeping the administration and management of the program as flexible as the law would allow. They took pride in thwarting the attempts of outside bureaucratic intrusion. The tight focus on program success and single mindedness of purpose was useful in warding off strict adherence to restrictive rules and regulations. A former NASA Program Associate Administrator recalls leasing aircraft during the Apollo Program to travel to important meetings, an action that he laughingly supposes would today land him in jail or at least guarantee a visit from the Inspector General. During Apollo the administrative staffs' main job was to keep the bureaucrats off the backs of program management and to provide whatever support was needed.

An understanding of the destructive effects of bureaucracy on post-Apollo NASA and the low cost effort can be found in works of three writers. Anthony Downs' book *Inside Bureaucracy*, Howard McCurdy's book *Inside NASA*, and Tom Alexander's article "Why Bureaucracy Keeps Growing" in *Fortune*.

Downs, now a Senior Fellow at the Brookings Institution, deals with the development of a theory of bureaucratic decision making. He postulates that bureaucratic officials seek to attain their goals rationally and are significantly—though not solely—motivated by their own self-interest. They tend to use whatever resources they control for their personal welfare. If Downs is correct, the low cost effort was

Administrator Blames Bureaucracy for NASA Waste

Wes Huntress, associate administrator for space science at NASA, said the agencies' own bureaucracy has resisted improvements that would dramatically lower operation, costs, freeing NASA to run more satellites within a flat budget.

Space News—April 18-24, 1994, p. 10.

in trouble from the start. And "every organization's social functions strongly influence its internal structure and vice versa."

In a bureaucracy, Downs postulated, growth improves the chance of promotion. This certainly describes the NASA experience during the Apollo era. Many young engineers and managers earned the nickname "rocketeers," not for their expertise in building and launching rockets, but for their rapid rise through the civil service ranks. Even as NASA leaders were busy fighting off the external bureaucracy, they were busy building their own. During this period of growth, staffs grew large and then larger still. Organizations grew into multilevel structures with chains of command reminiscent of a military hierarchy. Information flow up and down the structures was encouraged, and soon NASA had layers of management reporting to other layers of management, with spin doctors at every level. More and more NASA employees were doing less and less mission work and more of the bureaucratic and institutional support jobs. Downs describes this phenomenon: "A very significant portion of all the activity being carried out is completely unrelated to its formal goal or even the goal of the topmost officials." NASA evolved by the mid-1970's into the organization described by Tischler as one where large numbers of people spent most of their time doing things of little use to NASA's main mission.

Also, post-Apollo NASA was subjected to much more scrutiny on the part of Congress, the OMB, the White House staffs and the public. The honeymoon was over. NASA was just one more Federal government organization, and a lower level one at that, lacking the prestige and influence of cabinet rank. Real consequences of this fall from most-favored status involved more inquiries and oversight. And again, NASA's response proved one of Downs' most interesting postulation, that "any increase in the number of persons monitoring a given bureau will normally evoke an even larger increase in the number of employees assigned to deal with the monitors. This occurs because records can be read much faster than they can be compiled. To keep an additional monitor busy, the operating bureaucrat assigns two or more people to produce the report he demands."[3]

A solid indication of NASA's evolution into a formidable bureaucracy can be seen in the number of employees categorized as professional administrators, the category most likely to be involved in bureaucratic make-work. This number grew from 13.4% of the NASA work force in 1967 to 18.8% in 1991, an increasing percentage of a decreasing total population. And perhaps more surprising, post-Apollo professional administrators were more likely to win a promotion over their science and engineering colleagues, a sure sign of the diminishing importance of the NASA programs, the ascendency of the bureaucracy and the preservation of the NASA institution.

Tom Alexander tells us that "a bureaucrat seldom gains any personal reward from saving the taxpayer's money." In fact, the incentives tend to run in the opposite direction; the less money an Agency spends this year, the less it is likely to get next year.[4]

Saving money may have been a noble goal for post-Apollo NASA and may have made George Low's day, but it did nothing particularly good for a NASA employee, the employee's unit, or the parent organization. It was something the boss and the boss's boss simply were not interested in. Even if the NASA work force did not have the Apollo culture to contend with, it would have been difficult to sell cost reduction when there was an inherent resistance from the bureaucracy. Low's friend expressed it rather nicely: "There's nothing in it for me." NASA strengthens and nurtures its organizations by bringing money in—not by turning it back to the Treasury.

Downs also answers the question: How does a bureaucracy wind down or stop growing? He explains that by his theory, "Brakes on Acceleration."

The growth acceleration soon runs into serious obstacles. First, even though the bureau's original social function expanded greatly in relative importance, that function must still compete allocationally with others for social attention and resources. Therefore, as the accelerating bureau grows larger, it encounters more and more resistance to further relative growth of this function at the expense of other activities in society.

Indeed, this braking system did apply to NASA, just as Downs observed. NASA's peak in both budget and personnel came well before the culmination of the Apollo Program. Downs mentions three more "brakes" to the acceleration of a government Agency. "Second, the ever-expanding bureau soon engenders hostility and antagonism from functionally competitive bureaus." The sponsor of Downs' RAND study, the U.S. Air Force, was one existing bureaucracy that wondered about its own role in aerospace with the rise of a civilian space Agency. "Third, the bureau encounters the difficulty of continuing to produce impressive results as its organization grows larger and unwieldy." Without such results, the Agency can only get support from its suppliers, in NASA's case the aerospace industry. Finally, Downs points to internal checks to growth in addition to the social functions. "Fourth, conflicts among the climbers who flood into a fast-growing bureau provide an internal check . . . A high proportion of their efforts is devoted to internal politics and rivalry rather than performance." The major rivalries in NASA became Center vs. Centers vs. Headquarters. Thus, in Downs' scheme, the conflict among NASA Centers leads to a braking on accelerated growth.

Government bureaus also wane for other reasons. Downs describes "the age lump," a phenomenon that occurs when the Agency's top leadership consists of older people at about the same age or rank. A "decelerator effect occurs when the bureau is forced to reduce its total membership because of a sharp drop in the relative significance of its social function." Agencies die when their social functions are no longer desired or needed and also when the functions are valuable, but the Agency is unable or unwilling to perform them adequately or when they are taken over by a bigger and better Agency. But few agencies ever die: "the ability of bureaus to outlive their real usefulness is part of the mythology of bureaucracy," says Downs.

However, the main issue for George Low and the Low Cost Systems Office in the 1970's was not the management of growth but rather the process of change.

It's Easier to Kill a Vampire

According to Ralph DeGennaro, it still is easier to kill a vampire than an unneeded government program such as NASA's Advanced Solid Rocket Motor.

Killing a vampire only requires driving an ash wood stake through the vampire's heart, murmuring secret incantations, and burying him face down near a crossroads at midnight when there is no moon.

Space News— September 27, 1993.

Pushing Change at NASA

"Change is painful," NASA Administrator told the Wall Street Journal in the Aug. 30, 1993, issue. "My job is not to make people comfortable." Goldin was said to be working with NASA culture when he asked for faster, cheaper, smaller missions to replace big, lumbering projects that take a decade to deliver science.

How does an Agency or bureaucracy change its way, renew itself, improve? Once again, Anthony Downs has a perspective showing how agencies can react to what he calls "the performance gap." Termination or completion of a finite task, such as the Apollo Program, was NASA's big performance gap, just as the discovery of a polio vaccine produced a gap for the March of Dimes.

Enter the forces of inertia. An unbiased observer would think that NASA would be forced to change after Apollo. But the forces working against change can be more powerful, Downs claims. His proposition: "the larger the costs of getting an organization to adopt a new behavior pattern, the greater will be the organization's resistance to it, other things being equal." In other words, with so much at stake for NASA, strong resistance set in and took hold.

Downs notes two corollaries to this resistance to change:

- Each official's resistance to a given change will be greater the more significant the required shift in his behavior, that is, the "deeper" the layers in his goal structure affected by the change.

- The more officials affected, the greater will be the resistance to significant change.

In other words, a large Agency will respond more to the forces of inertia than to the forces of change, despite (or because of) external factors or forces. Some call this "entrenchment" or "ossification," but the interesting notions are the depth and breadth of resistance to change. Downs draws a conclusion from this phenomenon:

- The larger the organization, the more resistant it will be to adopt any given change.

- Small bureaus tend to be more flexible.

The fact that a large bureaucracy resists change is hardly a new concept, but remains an important one. Previous attempts to discipline the government bureaucracy have not fared well. The President's Commission on Economy and Efficiency was set up in 1910. It stressed performance, not status quo. The President's Commission on Administrative Management, also known as the Brownlow Committee, was formed in the Great Depression, but it led to the proliferation of Federal Agencies. The famed Grace Commission of the Reagan Era produced a 21,000-page report on government waste and inefficiencies, plus 1.5 million pages of appendices and documentation. Nothing happened. Vice President Albert Gore more recently launched the National Performance Review, the latest effort to discipline the agencies. The success or failure of this effort may very well depend upon 1) whether the theories of Anthony Downs hold true once again, and 2) how well we learn from past efforts to reform the bureaucracy, including the lessons learned from NASA's Low Cost Systems Office.

As if self-interest were not enough to contend with, cost reducers have to do battle with the Congress. Tom Alexander writes that Congress is as guilty, if not more so, as the bureaus of engaging in government expansion and consequent waste. In the case of NASA, this turned out to be not so much expansion as anticontraction. It is rumored that in responding to special interest constituents, powerful members of the Senate blocked the Administration's and NASA's plans to reduce its size by giving portions of a Field Center to the Department of Energy. This union of Congress, bureaus and special interests is known as "the iron triangle." The triangle works as follows: a member of Congress gets assigned to a committee dealing with prominent special interests in his or her district. The committee over time is dominated by Congressional members who favor "generous appropriations" for their special interests. In NASA's case, the best example of this is the Space Station Program, where billions of dollars have been spent with little to show for it. The money was spent in districts of powerful committee members effectively creating a quasi-jobs program or expensive welfare effort. From a low cost systems standpoint, going up against "the iron triangle" was a no-win proposition.

Berlew Revisited

In the intervening two decades much new research on culture has occurred and it is useful to apply this understanding to Berlew's findings.

From the beginning, few truly believed that NASA was serious about reducing cost. Berlew called this "a major disconnect." It is commonly accepted that most people look to behave in a way that is most likely to be rewarded. Despite this, there are many situations in organizations where what is rewarded (or perceived to be rewarded) is very different from the behaviors being promoted through management instructions, presentations, training and other assorted public forums. In the case of the LCSO, "a major disconnect" clearly existed between what managers heard being espoused and what they believed was being rewarded. In his report Berlew noted the lack of direction and inconsistency of the system as two of the main detriments to creating a cost-conscious work force. "There is no motivation problem at NASA" according to Berlew's study.[5] There was a desire and commitment to do what was right for the Agency. However, there was a widespread disbelief that the low cost approach being touted was truly supported by the management structure. As one of the participants in the study mentioned, "High cost but successful projects get criticized, but all is forgotten after launch." Another said, "There is no relationship between advancement and recognition in the Agency, and low cost performance."

The Berlew study indicates that the conflicting messages, inconsistent policy and actions, the perceived differences between what is being espoused and what is being rewarded, and the strong NASA culture were key factors in limiting the effectiveness of the goals of the Low Cost Systems Office. Considered from hindsight, these problems led to the eventual demise of the effort. It is now clear that four key factors in organizational functioning created the conditions for failure:

A CBO report raises questions about manned flights and questionable missions

The problem with the National Aeronautics and Space Administration these days is that, like some aging baby boomers, it's still stuck in the 1960s. It remains obsessed with manned space flights that cost an enormous amount of money and generally produce little of scientific value. It's no surprise then, that NASA is perpetually pleading poverty.

L. A. Daily News, Viewpoint—Sunday, March 27, 1994, p. 2.

Administrator Calls for Constant Change

NASA Administrator Dan Goldin, in Aerospace Daily for March 14, 1994, called for vision and evolution at NASA, not just programs that fight for survival.

unclear goals, poorly defined role expectations, inconsistent procedures for implementation and a strong negative culture.

The lack of goal clarity is evident in the remarks of the 24 senior staff people interviewed by Berlew. The relationship between lower cost and excellent performance was an obvious area of confusion and conflict. Many were concerned that the high emphasis on cost would lead to lower performance standards. One inconsistency was the fear of compromised performance—a fear that was abhorrent, even conceptually, when compared to the benefit of reduced cost.

A similar line of thought expressed in Berlew's report suggested that while a low cost philosophy could work, it would need to be coupled with increased risk. It was pointed out that no one ever accepted the idea of a possible increase in risk as a consequence of reducing cost. In fact, any failure or reduced reliability of effort could not be allowed; George Low simply would not accept any failure. Therefore, no open discussions could take place addressing the apparently inconsistent goals of lower cost and minimal risk. This lack of discussion (first breached at the Low Cost Workshop), coupled with confusion over the relationship between cost and technical performance, would become a major weakness of the low cost effort. The perception that low cost would invariably lead to compromised performance was the greatest criticism of the low cost effort.

Another area of weakness detected by the Berlew analysis was in explaining just how NASA would benefit from reducing costs. There was a strong suspicion that the goal of low cost was a public relations event designed to satisfy Congress and create an appearance of cost cutting. A typical comment: "NASA is dishonest; we've been forced into it by Congress." Low cost was considered just another "Washington publicity stunt" with few specifics and few actual benefits for NASA programs.

In terms of inconsistent goals, Berlew noted that NASA failed to define the roles and responsibilities needed to implement a low cost philosophy. Who champions this approach? Who would be responsible for implementing recommendations and making them part of the everyday culture? How would the NASA work force be prepared for these new cost reduction approaches? Who would identify the requirements for such a new approach? What skills would be needed by the work force to meet the changing requirements? How would changes be communicated to the larger organization, accustomed to working in a different environment?

It would be simple enough to claim that the culture was incapable of supporting the new way of doing business represented by the Low Cost Systems Office. However, such an assertion would not explain the few successes attributed to the effort. It seems that the ways in which individuals and organizations get motivated to accomplish new reforms were lacking in this effort. They include:

- The ability to deal firmly and swiftly with dramatic change.

- Clear program implementation plans and follow through.

186

• Definite expectations and reinforcement for low cost.

The broad-based effort of the Low Cost Systems Office promised radical shifts in the way business was conducted at NASA. Such dramatic change would invariably create tension and a certain degree of pain. Many issues involved in managing this change were poorly attacked or totally ignored. Two key issues were never addressed: Who was in charge of the program, and who were the targets of the program?

In his book, *Managing at the Speed of Change*, Daryl Conner identifies four roles critical in change: sponsors, agents, targets and advocates. Each role has an important part to play; however, if any of the roles are not effectively coordinated, the change is typically doomed to failure. Sponsors are sources of power and influence—individuals or teams with the ability to encourage large scale commitment to new approaches. The agents of change are those individuals who must assist in spreading and implementing the actual plan of action. The targets represent the groups or individuals who must accept, practice and ultimately support the new approach. Advocates are early supporters of the change who do not have the influence to make the change a reality on their own.

In terms of the Low Cost Systems Office, the roles for this major effort were never clearly defined. While there was a general understanding of roles of the known sponsors, advocates, agents and targets of change, it was never known who was expected to play which role. George Low was serious about the low cost effort, but it is evident that he did not energize a critical mass of advocates.

While change agents were given resources (a budget, office space, letters in support of the effort), they were not given a strong institutional mandate to proceed with the approach.

Furthermore, the numerous activities attempted by the LCSO made the effort look "a mile wide and a foot deep." As Conner states, "Major change will not occur unless the appropriate sponsors demonstrate the sufficient commitment."[6] George Low was a strong advocate of the Low Cost Systems Office at first, but he lacked the sustained commitment and unrelenting focus to be an effective sponsor of this vastly new effort.

The strong culture of NASA was indeed a barrier to the low cost approach. Without a doubt, the success of the Apollo mission made a new way of doing business an odd idea and an affront to a successful organization. While NASA was still celebrating its great success, Low and others outside of NASA were becoming concerned that NASA programs were too expensive and too lengthy to be able to compete effectively in a more cost-conscious world, especially with President Nixon being unsure of NASA's role.

The success of the Apollo program made it extremely difficult to generate the organizational dynamics necessary for NASA to change. Daryl Conner's work indicates that two of the most important prerequisites for change are pain and remedy. Pain is the catalyst for breaking away from the existing way of doing things. Unless an individual or organization experiences or expects to experience significant problems, there is always more support to maintain the status quo.

Kurt Lewin placed the concept of pain within the context of environmental forces and the force field analysis, which suggests that any situation is in a fluid state, with factors competing to maintain the status quo and other factors competing to create change. In order to manage any new change, it is necessary to create a sense of dissatisfaction with the status quo and at the same time offer a viable solution or remedy.

The low cost proposal was initiated at a time when few in NASA felt any pain. Apollo was recent enough to allow continued basking in the glory, and very few realized that the Space Shuttle program may well have saved the Agency from major cutbacks. The Low Cost Systems Office was also never firmly established as a remedy or solution to any specific problem. Even those potential supporters of an initiative that promised cheaper and shorter duration programs were never satisfied in seeing how the Low Cost Systems Office could help. In the interviews Berlew conducted with senior NASA personnel, the lack of perceived benefit or remedy was evident. As one individual stated, "Members are told to think and behave in new ways, but get confusing, often conflicting messages about what those new ways are." Even those who realized that the old ways of doing business might no longer be appropriate found the low cost alternatives unclear or unattractive.

Another key stumbling block was the lack of support from the Directors of each NASA Center. In the highly decentralized structure of NASA, lack of support from the Directors who ran the Field Centers would doom any initiative. Although the Low Cost Systems Office seemed to have some early support from a few Center Directors, even their interest waned. When their support was necessary in as simple a thing as establishing Center-located low cost systems offices, their efforts were hardly inspiring. The personnel assigned to the Center low cost offices were typically looked upon as unworthy of more meaningful responsibilities at the Center. The selection of such candidates only emphasized the disinterest of powerful potential sponsors of the low cost system effort. The only sponsor was George Low, and in a culture which empowered the Center Director to the greatest degree, that proved to not be enough.

The lack of pain and remedy was also evident in the inadequate plans for low cost systems. In his change model, Richard Beckhard points out the need for thorough planning in each of the critical phases of a change—the present, transition, and future or desired state. It has already been pointed out that a clear common vision for the future of the low cost approach was never finely articulated. Similarly, a clear, convincing purpose of the office never materialized to gain the necessary widespread support. It should come as no surprise, therefore, that there was also no established plan for how the low cost approach would transition from the present to the future. It seems that the philosophy was to attempt independent test cases supportive of the low cost strategy and hope that they would provide the impetus for ongoing grassroots effort. Unfortunately, the lack of a clear road map only served to heighten the confusion and increase the anxiety about what the Low Cost Systems Office was trying to accomplish.

Another major barrier to systematic cost consciousness seemed to be a fundamental tendency of human behavior: *What's in it for me?* In the case of the Low Cost Systems Office, the rewards were never established to encourage any kind of advocacy. In fact, most individuals alluded to their belief that the espoused commitment to low cost by senior NASA management was meaningless. Many mentioned the reality that concern for cost was more of a detriment to personal goals than a facilitator of career aspirations. The low cost approach led to the absurd situation where individuals would claim to support the office, but behave in ways that led to contrary actions.

The work of Leon Festinger seems to connect appropriately with some of the seemingly contradictory opinions and actions that were expressed at that time. Festinger described the notion of "cognitive dissonance," a term he used to describe a situation where a person has two contradictory cognitions (ideas, opinions, or beliefs) about the same thing.[7] For example, an overweight person may like eating fatty foods and at the same time start to get inundated with evidence that links such foods with obesity and greater health risk. One way to reduce the dissonance would be to downplay or disagree with the evidence that suggests greater health problems due to such a diet. In the case of the Low Cost Systems Office, it seems that there was support and evidence which pointed to the need and benefits of such an approach. This support was indicated by the Berlew study. At the same time, individuals confronted with a new way of doing business and maintaining the more comfortable status quo of the present seemed to line up against change. Ironically, those who often cited the potential benefit of the low cost approach often were the most critical of the weaknesses of the office.

The lack of commitment of the Low Cost Systems Office can also be attributed to an ignorance of the factors that motivate people to behave in a low cost manner. Nadler and Lawler indicate that, according to expectancy theory of motivation, individuals will be motivated to behave in a particular way if new variables are in place:

1. The individuals believe that their behavior will be able to lead to the desired outcome.

2. The individuals believe that outcome will have a positive benefit (some reward) for them.

3. The individuals believe that they can perform at the necessary levels for success.[8]

The Low Cost Systems Office was never quite able to let people know how their behavior could lead to the desired vision of the office. A more fundamental problem was that the office could never articulate a clear vision of the low cost approach and the potential benefits the Agency and its members would reap from its implementation and success. As a result, the effort collapsed.

A common phenomenon in the attempt to establish a low cost environment was the apparent disconnect between beliefs and behaviors. Many key people were actively involved in developing and supporting activities which they did not believe would or should work. Why would individuals who did not believe in the Low Cost Systems Office expend so much energy in its behalf?

One answer lies in the theoretical work of Kurt Lewin and his followers. (See Deutsch and Krauss, 1965.)[9] Lewinian theory contends that there is a state of tension within a person whenever a need exists. As such, people attempt to alter their environment (move toward goals) in a way which reduces the tension. In cases where a goal is clear, the individual moves toward it; in cases where it is unclear how to reduce the tension, the individual still searches for the appropriate goals.

An extension of Lewin's theory is found in Leon Festinger's theory of social comparison, which helps to explain why people will behave in ways that are apparently opposite their true or initial beliefs.

Social comparison theory makes basically three assumptions. First, people want to find out if their beliefs are correct. Second, people will seek out others like themselves to serve as the reference point. (Center Directors, for example, looked to other Center Directors for opinions on the Low Cost Systems Office.) Third, people will change their behaviors to fit with their referent group. (Thus, managers who do not really believe in low cost systems will become more supportive if their peers appear to be supportive.)

Festinger's cognitive dissonance theory expands this notion of behavior change. People will change their behavior, thoughts or feelings in order to be consistent with their other behaviors. This theory is critical in pointing out the importance of referent groups to motivate behavior change, and also in indicating how people attempt to make reality, behavior, and belief consistent.

These factors help to reconcile the inconsistency between the expressed beliefs in low cost and the support depicted in activity and behavior. A manager who had little faith in the effort would not be expected to be involved in supportive activities. However, if the manager were aware of the Administrator's support for the effort and witnessed the surrounding activity, a tension would exist between individual beliefs and the need to appear supportive of the larger effort. Somehow the manager would need to find ways to be supportive and still be detached in order to be prepared for the inevitable failure of the initiative.

The Low Cost Systems Office seemed to be plagued with a quiet sense that it could not be successful. Despite the activities in support of the effort, the lack of belief in the initiative would prove to be fatal. Sufficient belief had not been built up to guarantee ongoing support.

What are the implications for bringing on any new major change initiative? How could the Low Cost Systems Office have been better leveraged to ensure ongoing success? The weakness of the initiative was tied to the inadequate attention paid to understanding and believing in the importance of low cost. Much is written about the need for such major changes to be accompanied by a strategy which is suited to exploiting and accommodating an organization's existing system. If such an approach does not work, nothing short of changes in key personnel could assist a major change.

The writings of Katz and Kahn (1978)[10] suggest that leaders must be able to understand and respond to the requirements of the organization's perspective. A systemic perspective calls for the ability to adapt to the internal and external situations. At the same time, they emphasize the importance of leaders to distribute or share their leadership. They write, "By and large, those organizations (and successful leaders) in which influential acts are widely shared are the most effective. The reasons for this are in part motivational, having to do with implementation of decisions, and in part nonmotivational, having to do with the excellence of decisions." As they explain the success of distributive leadership, Katz and Kahn provide specific examples of managers who voluntarily modify formal organizational structure in order to provide employees with true influence in decisions, access to relevant information, and delegated responsibilities.

Low's Departure

We have covered the most probable reasons for the collapse of the low cost effort. All of the aforementioned were significant factors in its ultimate demise, but the smoking gun, the one that fired the fatal shot, may well have been held by none other than the founder of the program, George Low. Low was a brilliant engineer and the program manager widely credited with saving the Apollo Program after the tragic launch pad fire that claimed the lives of three astronauts. He had been with NASA from the day of its founding. He helped design and manage the organization that successfully accomplished one of the greatest engineering accomplishments of all time. He was proud of NASA as a unique "can do" organization, and he eagerly looked forward to new challenges. However, as the marketing of the Shuttle program was to demonstrate, he was not above "buying in" if it meant the survival of NASA as the Federal civilian space Agency.

The late Richard Feynman, a Nobel laureate and member of the Challenger Investigation Group, later reflected on NASA's and Low's decision to build the Shuttle.

> When NASA was trying to go to the Moon, it was a goal that everyone was eager to achieve. Everybody was cooperating, much like the efforts to build the first atomic bomb at Los Alamos. There was no problem between the management and the other people, because they were all trying to do the same thing. But then after going to the Moon, NASA had all these people together, all these institutions and so on. You don't want to fire people and send them out in the street when you're done. So the problem is what to do.
>
> You have to convince Congress that there exists a project this organization can do. In order to do so, it is necessary (at least it was apparently necessary in this case) to exaggerate how economical the Shuttle was

191

going to be, to exaggerate the big scientific facts that would be discovered. In every newspaper article about the Shuttle there was a statement about the useful zero-gravity experiments—such as making pharmaceuticals, new alloys and so on—on board, but I've never seen in any science article any results of anything that have ever come out of any of those science experiments which were so important! So NASA exaggerated how little the Shuttle would cost, they exaggerated how often it could fly to such a pitch that it was obviously incorrect—obvious enough that all kinds of organizations were writing reports, trying to get the Congress to wake up to the fact that NASA's claims weren't true.

I believe that what happened was—remember, this is only a theory, because I tell you, people don't agree that although the engineers down in the works knew NASA's claims were impossible, and the guys at the top knew that somehow they had exaggerated, the guys at the top didn't want to hear that they had exaggerated. They didn't want to hear about the difficulties of the engineers—the fact that the Shuttle can't fly so often, the fact that it might not work and so on. It's better if they don't hear it, so they can be much more "honest" when they're trying to get Congress to OK their projects.[11]

About the time George Low made his first low cost speech it was obvious to some that NASA would not have a Shuttle that would even come close to providing easy and inexpensive access to space. It was also apparent that the mission model used to justify the economics of the Shuttle was a NASA-generated fantasy. Even with these difficulties, Low initiated the low cost effort. Granted, there were and still are many opportunities to reduce costs in NASA, but the big savings that would be necessary to fulfill Low's vision of cheap access to space were not to be found anywhere. No launcher capable of providing such access to space was going to be available in his lifetime, and Low was not inclined to voluntarily shrink the NASA organization he helped found. So he settled for a much reduced and limited cost reduction effort, one that he hoped would produce some solid results but one that would not revolutionize the space business. As critics knew, at times it was more important to "look" low cost than "be" low cost.

Del Tischler said he left because he was convinced Low was not vigorously supporting the program. In Tischler's words, Low wanted a catechism, Tischler wanted reformation. Tischler and most NASA managers questioned Low's placement of the newly formed Low Cost Systems Office under the Associate Administrator for Technology, Utilization and Industry Affairs, a backwater office of little importance in the day-to-day workings of NASA. Worse yet, the head of this office, Ed Gray, was not considered to be a member of Low's trusted inner circle. Even close confidants of Low's were puzzled by this organizational move. It did not fit Low's rhetoric, and it sent a confusing message to the NASA work force.

Dr. Robert Cooper was recruited in the spring of 1976 by Low for the position of Director of the Goddard Space Flight Center at the height of LCSO activity. In his pre-employment discussions with Low of the future of NASA, Dr. Cooper does not recall discussing Low's vision of a new and less costly NASA, a curious omission. Low also imposed no consequences for not participating in the low cost program. It appeared acceptable to watch from the sidelines and simply shout amen at the appropriate time. Low punished no one for not achieving lower costs. The low cost workshop was thought by Muinch and Gray to be a major event in renewing the program momentum, but Low did not insist that the Deputy Center Directors or other important NASA officials attend, and few did. He did not chair the sessions and only made cursory appearances at the beginning and end of the workshop. At this crucial time in the conduct of the low cost program, George Low displayed a surprising disregard of his creation. When George Muinch accepted Low's offer to head up the LCSO, he did so with the view that "if lower cost approaches were really going to be promoted and supported at the NASA Administrator level, then I could make a contribution. . . . I also recognized that change would be very much opposed by many of the career NASA managers and staff. Top management support from within NASA was, of course, the key to dealing with this opposition." Muinch today believes the Low Cost System Office did have an impact on the NASA culture but believed more could have been accomplished if Low had not left NASA. Once Low left, the Centers knew they were free to disregard the LCSO.

George Muinch felt cost improvements had been accomplished, but NASA management had a strong mindset: "NASA programs can succeed as reliable and safe systems only when the Centers are completely free to do things as they always had. Lower cost efforts do not have a place on our priority list or at best they are at the bottom of the list. Cost effectiveness efforts imposed on the Centers are fruitless and counterproductive."[12] Muinch, too, believed that without top management support the office would be totally ineffective.

In reviewing the tribulations of the Low Cost Systems Office, it seems that Deputy Administrator George Low never really fully supported the effort. It appeared that he cranked up the LCSO, gave it a spin and walked away. We can only speculate why he would have given up, why he abandoned the low cost effort. Had he pushed the reforms of Del Tischler or had the Agency backed George Muinch in his efforts to create a cheaper, better, faster NASA, the resulting Agency would certainly not resemble the NASA of Low's glory days, the NASA of Apollo. But then, post-Apollo NASA did not resemble the vibrant, exciting NASA of the 1960's, either.

It may have been that George Low, in his wisdom, had decided to go through the motions of reducing cost in order to appease a distracted President and a disinterested Congress. Did he know that NASA was not ready to downsize, reinvent itself, become reengineered? If so, he never told anyone else, but he launched the effort, let it take its own course and then left before it collapsed.

In so doing, George Low may have left behind his greatest legacy: a blueprint for a cost-effective NASA, whenever it might be ready; a roadway for cutting and reducing costs. In other words, the Low Cost Systems Office may have been an idea whose time had not yet come. Had it been forced on NASA at that time, low cost systems may not have worked anyway. Maybe George Low knew that. Maybe he also knew he was building, documenting and leaving the prescriptions for an ailing Agency of the future, an Agency with a greatly reduced budget and major cuts to the work force. Some say that future is now.

Footnotes

[1] John P. Kotter and James L. Heskett, Corporate Culture and Performance. New York, Free Press, 1992.

[2] Humboldt C. Mandell Jr., "The Practical Side of Implementing Cultural Change in NASA" in a letter to the author, August 3, 1994.

[3] Anthony Downs, Inside Bureaucracy: A Rand Corporation Study. Little, Brown & Co., Boston, 1967.

[4] Tom Alexander, "Why Bureaucracy Keeps Growing," Fortune, May 7, 1979.

[5] David E. Berlew, "Creating A Low Cost Culture Within NASA," Development Research Associates, Cambridge, Mass., October 30, 1975. Order for Supplies and Services No. W-13, 872.

[6] Daryl R. Connor, Managing at the Speed of Change. New York, Villard Books, 1993.

[7] Leon Festinger, A Theory of Cognitive Dissonance. Stanford, Calif., Stanford U. Press, 1957.

[8] D. A. Nadler and E. E. Lawler, "Motivation: A Diagnostic Approach," in Perspectives on Behavior in Organizations, edited by J. R. Hackman, E. E. Lawler and L. W. Porter. New York: McGraw-Hill, 1977.

[9] Morton Deutsh and Robert Krauss, Theories in Social Psychology, New York, Basic Books, 1965.

[10] Daniel Katz and Robert L. Kahn, The Social Psychology of Organizations, New York, John Wiley and Sons, Inc., 1978.

[11] Richard P. Feyman, "An Outsider's View of the Challenger Inquiry" in Physics Today, February 1988, p. 37. With permission.

[12] George Muinch, letter to the author, September 20, 1994.

Chapter 14

Lessons Learned, Ignored and Forgotten

The impact of the Low Cost Systems effort on the NASA organization, its culture and the way it conducted business, was minimal. Costs were reduced, but not substantially; standard equipment and multimission spacecraft were used, but not to the extent hoped; and proven cost control techniques did not replace the free spending habits left over from the Apollo culture. It can be assumed that with full management support and adherence to the blueprint developed by LCSO, much more could have been accomplished. But big change did not occur—there was no cheap access to space, and NASA remained an institution staffed and organized to support yet another Apollo-style program—hardly a low cost mode.

Tischler was right when he told Low that what NASA had to do to become a low cost provider of space services was not mysterious. In their first year of operations Tischler and his working groups had identified just about all of the aerospace industry's major cost drivers and the costs peculiar to NASA's management style. As a result they knew it would take much more than just tinkering with hardware to change NASA. Significant structural change was in order. It was a big job, but not a mysterious one.

What is mysterious and perplexing, however, is that many if not all of the cost issues identified and validated in the early 70's still plagued NASA and its supporting aerospace industry in the mid-90's. This inability to bring about change is explained by Richard Luecke in *Scuttle Your Ships Before Advancing*. Luecke tells of Robert Hayes at the Harvard Business School who observes that we "seem to be forever rediscovering the truth known to those that lived two generations earlier." Hayes mentions that he thought of calling this discovery the Hayes Law of Circular Progress until a friend made him aware of an 1843 edition of the Edinburgh Review in which it was written: "In the pure and in the Physical Sciences, each generation inherits the conquests made by its predecessors . . . But in the moral sciences . . . particularly the arts of administration . . . the ground seems never to be incontestably won." Luecke observes

> that each generation receives the hard sciences intact, but must ever
> relearn the moral sciences explains, perhaps, why we have succeeded in
> putting people on the moon while failing to resolve fundamental problems
> in education, employment and international peace. . . Today, our ability to
> lead others in important undertakings, to manage organizations, to deal

with human conflict and change, and to avoid repeating the foolish mistakes of our predecessors is probably no greater than was that of people who lived hundreds—or thousands—of years ago. While science and medicine race forward, the people-related disciplines seem to go around in circles. The reason for this disappointing state of affairs may be the fact that the skills of leadership, of managing, of interpersonal relations are not easily taught in textbooks or in the classroom, but must be gained instead by individuals through their own experience."[1]

NASA has certainly suffered from this Law of Circular Progress and it has cost the institution dearly. At the completion of the Apollo Program, George Low and others documented their lessons learned. They said the single most important reason for that program's success was the simplicity of the technical and managerial interfaces. In other words, the juncture where one Center's work left off and another's began was purposefully kept as simple as practical. The interface between the Apollo command and service module, for example, managed by the Johnson Space Center, and the Saturn launch vehicle managed by the Marshall Space Flight Center, consisted of just nine wire bundles. With that degree of simplicity Low was comfortable with just one NASA employee looking after it. He postulated that if there had been a hundred bundles, as many as a thousand civil servants and contractors could well have been involved in managing this interface.

Twenty years later, the Space Station program's management scheme included managerial and technical interfaces as complex and convoluted as NASA or industry had ever experienced. In just two decades, the most important lesson in managing the world's most complex program was either forgotten or never learned by the new team of NASA managers. This disregard of past experience still plagues the Space Station program. As Gilbert Roth, the staff director of NASA's Aerospace Safety Advisory Panel, said in an address to the 1990 Space Transportation Propulsion Technology Symposium held at Penn State University, "The greatest lesson we seem to learn is that we seldom learn from lessons learned."[2]

There are plenty of lessons to be learned in the Low Cost Systems exercise. Following is a descriptive list of some of the most important of the macro class lessons learned.

Understand the Role of the Founder as Change Agent. NASA's experience would indicate that founders make poor change agents, especially when the change is of a drastic nature. It's difficult to maim that which you create. George Low was as emotionally tied to the Apollo NASA as humanly possible. It was his organization and people who did the impossible on time and within budget. At the time, NASA represented the highest level of human achievement, an achievement not measured in profit or loss but in the triumph of the human spirit. NASA of this era was a true emotional experience. You had to be there to fully understand the dedication and joy most employees experienced in participating in mankind's greatest feat of exploration. How could a founder of such an organization dismember it? The answer was he could not, and there were literally thousands of George Lows in NASA. Perhaps change in such an organization can only be successful after all the founders have, one way or another, left.

Establish Meaningful Incentives. As Anthony Downs postulated and the LCSO experience demonstrated, there are no incentives in a bureaucracy to manage in a low cost manner. In fact, just the opposite is true; there are plenty of incentives to overspend. It is counterproductive for an individual or organization not to spend all of the approved budget or not to build the largest structure possible to support its own growth, income and power. Those who have money left over at the end of one budget cycle will invariably suffer a cut in the next budget. Until meaningful incentives for savings can be established and implemented, future low cost efforts are doomed to a similar fate of NASA's Low Cost System Office. Remember what Low's friend said: "There's nothing in it for me."

Share the Vision. Low, for all his initial enthusiasm and drive, never convinced even his closest associates of the urgent need to change NASA, let alone to support a low cost program. They never bought into his vision of a cost-effective agency serving a broad base of new customers. Without this shared vision and top management support, the Low Cost Systems effort stood alone, opposed by field Center managers and Headquarters program directors

Know the Culture. Attempting to alter long term organizational customs and beliefs without a detailed knowledge of what they are and how they should be addressed is sheer folly. It wasn't until Berlew's report spelled out in startling clarity the dimension of some of the cultural obstacles facing NASA's low cost initiative that the complexity of the task was understood. Challenging the culture of the NASA work force was not as neat and clean, nor nearly as much fun, as tinkering with hardware.

Invest in the Work Force. The low cost effort revealed something quite unexpected—the importance of people over hardware and process in controlling cost. Manager after manager agreed that the most important element in cost control is a highly motivated work force. The ideal work force would have many of the attributes of those who managed Apollo: personal desire to excel, self-organization and direction, a shared vision of the program and its goals and objectives including total costs and an almost fanatic desire for mission success. The managers of such a work force must be experienced and have the authority and responsibility to get the job done. Developing this work force is expensive and time consuming but absolutely essential to low cost performance.

Keep Organizations Lean. Unique programs with a well-defined beginning and end and that can be measured in terms of performance, cost and schedule should be organized and managed in a project manner. This type of focus inhibits the formation of large bureaucratic support organizations that all too often remain in place long after their reason for existence has passed. Organizational flexibility is needed to ensure good management and cost control in the project mode.

Assign New Initiatives to Line Organizations. Most industry and government organizations take the worst possible approach in attempting to implement a new initiative such as the Low Cost Systems program. They invariably organize a small headquarters staff office, reporting to the CEO or other important bureau or corporate offices, and charge this staff with implementing the new management process. Immediately, battle lines are drawn between the large and robust line organizations, who will eventually be asked to implement some portion of the new program, and the smaller staff office whose only clout is, "But the boss said. . . ." The line usually has no vested interest in the new initiative; the responsibility for its success, after all, rests with the headquarters staff office. Conflict and chaos follow, with the staff office on the defensive without real authority. The new initiative never has a chance to be "institutionalized." In the case of the LCSO, the all powerful NASA field Centers opposed it from the beginning. After five years the Centers won; Low left and the office was closed. A much better approach would be to initiate the new process as a pilot operation, with the highest priority, in the line organization. If it succeeds, ground rules and guidelines for implementation across the entire organization can be developed.

Standardize Equipment. Equipment standardization works. It works for spacecraft and automobiles. It works so well that some of the original 22 items of standard equipment sponsored by the LCSO effort in the mid-70's are still flying. That in and of itself is an incredible feat, and partially puts to rest two major objections to standardization: first, that standards will stultify technology, and second, that standards will have an extremely short lifetime—much less than the five years predicted by the LCSO.

Rather than debate the efficacy of standardization, the discussions now should focus on how to make it work today. An obvious approach would be standardizing at the form, fit and function level, similar to today's computer industry. Future standardization efforts should include global requirements, not just those of NASA or DoD missions. Such efforts could save the world's spacefaring nations millions if not billions in development costs.

Tracking and cataloging available flight proven equipment would also pay enormous dividends. Time and again, off-the-shelf is cited as an effective approach to reducing cost. Project Clementine, a recent successful lunar mapping mission, is touted as a low cost success. The project, sponsored jointly by the Department of Defense and NASA, proved a number of the LCSO contentions regarding standard and available equipment. As far back as 1972, Tischler's Low Cost Task Force came to the conclusion that "aerospace managers must become attuned to assembling space systems in contrast to developing them." This admonition, which implies the use of standard or available equipment, was one of the most important elements in keeping Clementine's cost in control.

Identify and Address Culture-Driven Costs. Tischler was right again—NASA's management culture was much more important to cost performance than standardizing hardware, and the program and business practices effort was especially effective in identifying major culture-driven costs. Unfortunately, little progress was made in addressing them then, and, consequently, many still plague today's aerospace programs. These lessons were not learned. The most frequently identified of these are:

Buy-ins—A buy-in by industry is defined as "the submission of an offer, usually substantially below estimated cost, with the expectation of winning the contract."[3] For government, it's a deliberate underestimation of expected costs to win approval of a proposed new project from agency management, the Administration, and the Congress. This problem is so extensive it prompted the National Academy of Public Administration to report:

> Regrettably, the willingness to recognize the magnitude of funding requirements for space programs... has been uncharacteristic of much planning for U.S. civil space activity. Far more common has been the tendency to underestimate costs, as in the case of the original estimates for the Space Station and other large projects such as the Hubble Space Telescope and NASA's Tracking and Data Relay System. These estimates proved to be based on unrealistic assumptions deriving from best case scenarios. They succeeded in obtaining the official authority to proceed, but they have presented many serious budgetary problems in the implementation of the programs.[4]

Buy-ins are not a recent phenomenon. Thucydides remarked 2400 years ago: "Their judgement was based more on wishful thinking than on sound calculation of probabilities; for the usual thing among men is that when they want something, they will, without any reflection, leave that to hope, while they will employ the full force of reason in rejecting what they find unpalatable." And in 1913, Frederic Haskin, in writing about the building of the Panama Canal, commented: "The early estimates made by the American Engineers were far too low, but the French experience had taught the United States to expect such an outcome. Indeed, it is doubtful if anybody believed that the first estimates would not be doubled or quadrupled before the canal was finished."[5]

Why deliberately underestimate? Simple. . . a low estimate stands a better chance of being accepted by all parties than a realistic one. As long as there are no sanctions for NASA, its contractors and, on occasion, even the Congress and the Administration to underestimate costs, buy-ins will continue to plague the government's ability to control cost. From day one, such an afflicted program suffers from the lack of a firm baseline, and for all practical purposes is totally uncontrollable. NASA buys in because this time-honored way to sell programs rarely results in cancellation, let alone recrimination, and NASA and its contractors can always "get well" in the following year's budget. Industry routinely buys in because of the federal government's "low bidder" procurement policy and practices, and in most cases (usually because of poor definition) industry knows that the government-initiated contract changes will eventually ensure the contractor an acceptable funding level. Congress tends to tolerate buy-ins because voters want the jobs and the economic well-being such projects bring to their districts.

Risk Policy—Much has already been said about the lack of a sensible risk policy and the accompanying effects on cost. The issue, however, may be endemic to doing business in space. Early in the low cost investigation it was pointed out by NASA project managers that no one, not NASA, the customer, the public or certainly the Congress, would be especially thrilled about a series of low cost failures. That observation applies equally to today's faster, better, cheaper missions. Then as now, no failure is acceptable. Space exploration and operations are still so dramatic and exciting as to preclude failure in the minds of many, and the costs of access to space are so high that failures reflect badly on the customer, sponsor and government alike.

NASA managers, its contractors and customers have spent billions attempting to avoid risk, thereby doing their part to keep the cost of doing business in space enormously expensive. Cheap, routine access to space would do much to solve this dilemma. The real key to a rational risk policy may be to couple low launch costs with a guaranteed opportunity to refly transportation failures.

Inadequate Definition—This topic would make any top ten list of issues in managing complex projects. It is a recurring problem frequently tied to the practice of buying in. Once a program is sold, there is a rush to get it under contract and begin "bending metal" as soon as possible. Later, when the real difficulties and associated costs become known, there is usually no option but to proceed, for the cost of contract termination is so horrendous it simply becomes less expensive to find additional funding. Improved project definition is imperative for overall cost control. In 1979, as the Low Cost Systems effort was terminating, NASA put together a study team to recommend changes to improve project management performance. The team reported its findings in 1980. A major recommendation was that:

> A NASA project should be well understood before it is approved for design and development. A thorough definition of the technical aspects, management (including the roles of the NASA Center), cost and schedule is required to estimate potential risks to NASA management, the Executive Branch, and the Congress as they contemplate approval. Up to 5 to 10 percent of the run out cost of a project should be expended during the definition phase. NASA managers must not assume that approval of definition funds automatically means approval and funding of the project itself.[6]

There are two principal reasons why inadequate definitions still occur. First, the lack of disciplined planning encourages large numbers of new start proposals; presumably the more new starts an organization has in the system, the better its chances of getting one approved. This gives rise to the second reason: since there are too many new start proposals in the system, there are too few dollars and in-house staff technicians to do an adequate definition on them all. The process then shifts from a technical to a marketing orientation—it becomes more important to sell than to define. In any event, there have been few consequences to the line organization for not doing an adequate job or few rewards for management to ensure they did. For low cost programs, it is essential that a thorough evaluation of technical and managerial alternatives be considered as an important element of project definition. The NASA experience is that major cost problems arise from poor definition, unrealistic requirements, and "requirements creep"—those expensive afterthoughts. More resources, both contractor and inhouse, need to be spent in the definition phase of the project cycle to identify not only where to begin but all the steps that bring it to completion.

Research and Development Within the Project Cycle—Frequently, and often due to inadequate definition, projects are allowed to proceed into the final design and construction phase with critical technology yet to be developed. What usually happens is the required technology development cannot meet the project schedule, the schedule slips, costs mount and the marching armies stand by while their cost clocks run on.

Relying on new technology to be ready to meet a project schedule is akin to the cartoon of two scientists looking at a very complex formula written on a blackboard and one assuring the other that at a particular point in the formula a miracle will occur. Recall the admonition of Tischler's working group that NASA and its contractors should concentrate on assembling, not developing, spacecraft payloads.

Courtesy of Maria Killingstad

Inadequate Cost Estimating—Although there may be a general improvement in the ability of government and industry to estimate costs, it is still at best a black art. NASA, other government agencies and perhaps the industry itself still do not know in advance what a complex project will eventually cost, or even why it costs what they ultimately pay.

Even experienced managers can be surprised at some costs. A high ranking MSFC official told the story of making a quick fix on the Skylab Project. In order to solve a viewing problem, it was necessary to drill three small holes in a plexiglass inner porthole. This job, he estimated, could be done by the average homeowner in about ten minutes with an electric drill and one drill bit. Despite his 30-plus years experience and familiarity with contract change procedures, the official was shocked to learn that this minor fix cost NASA $50,000. The time and materials for the job were minuscule; the real costs were associated with the NASA-imposed, contractor-managed paperwork systems procedures and change.

Cost estimating problems are compounded by the needs of the institution added to program development costs. Institutional costs, often referred to as wraps, include those associated with multi-organization participation, complex multiple interfaces, duplication of effort and facilities, and overhead costs. Unfortunately, few of these costs are included in the original project estimates. Even with the use of sophisticated computer-based models, cost estimating is still not done to the level of precision needed for exercising good cost control. In an industry dominated by engineers, who at times demonstrate a marked disregard for costs, much still needs to be done. Cost estimating must be recognized as a highly specialized technical profession on a par with any engineering discipline, and, as such, needs talented, well-trained people, tools and development opportunities. If NASA and industry are to reduce costs, they need the capability to identify, analyze and understand the things that drive program costs. Each cost overrun proves that time and again.

Unqualified Management—Many of the Apollo managers boasted of never having read a management book; nonetheless they did an outstanding job on Apollo and Skylab. There were two significant reasons for this. First, many of the "oldtimers" came from industry with considerable managerial experience. They joined NASA early and many left after Apollo. They were natural systems engineers and applied this discipline to the management challenges of the hardware programs. Second, adequate, or some would say more than adequate, resources made the Apollo management job relatively easy. However, post-Apollo NASA lost a number of these experienced systems engineering-oriented managers, and there was a noticeable reduction in resources available to manage the follow-on projects. Experienced managers will readily admit that it is much easier to manage with deep pockets.

Because of its good record during Apollo, NASA subsequently had little compunction in appointing managers with less than the demonstrated knowledge, skills or experience needed to manage multi-billion-dollar projects successfully. But the management job was infinitely more difficult in the era of tight budgets and many unskilled managers failed. Their projects were in turmoil, costs escalated out of control, but they usually had the good sense to quit before they were fired. The net results were projects in trouble and the losses of otherwise good engineers.

Contract Form—Successful definition efforts should enable any organization to use fixed-price contracts for much of their development activities. Fixed-price contracting would require a number of far reaching management changes: adequate definition, frozen requirements, a halt on research and development in the final design and development stages of the project cycle, and the use of previously developed designs and equipment. For NASA, it would mean altering its management role from partnering with industry in all development activities to giving industry a performance specification.

Cost savings demand new management and acquisition paradigms. Remember Low's visit to Fairchild where the staff said that NASA frequently got too involved at the subcontractor level, causing the prime contractor to lose control and allowing costs to grow? LCSO studies recommended NASA use more fixed-priced contracts or establish firm cost ceilings and let the performance come out where it would. The Fairchild employees also felt the Advanced Technology Satellite they built could have cost up to 25 percent less if NASA and the contractor had initially agreed to a realistic cost target and the importance of holding to that cost, and if NASA had not over specified. Or consider the way COMSAT, a commercial company, did business. Their satellites were procured to performance requirements only; details of design and production were left entirely to the selected contractor. All satellite hardware contracts were fixed price incentive fee, with the incentives based on performance and life span.

Unclear Goals and Objectives—Project managers must have a consistent set of unchanging mission goals and objectives backed by commitments from the institutional managers. The management team conducts meaningful trade-offs to performance, schedule and costs in meeting these objectives. Once the goals and objectives are agreed to, they are communicated to all project levels. Misunderstandings must be clarified and disagreements resolved before the project proceeds. The goals and objectives are then locked in and subject to change only to save the project. Managers must be delegated real authority and responsibility to carry out mission objectives. Portions of this authority cannot be retained by the next management level. There can be only one person in charge, the one fired if the project fails. Appointing good project managers, giving them a chance to run their programs and allowing them to remain in the position long enough to get the job done, are vital to controlling costs.

Development Time—A colleague recently remarked sadly that all his life his dear mother had been lying to him. Time and again she had reminded him that "haste makes waste." But now he was convinced that "haste saves cost." NASA's experience is that long development times not only increase costs, but also increase technical and political risk. The longer the development cycle, the greater the opportunity for NASA and its customers to change the technical requirements, further delaying the program, and allowing for the political climate to change. Time in development is directly correlated to the degree of completeness of the preceding definition process. Once the program is properly defined and its requirements and specifications prepared, the development will proceed with dispatch. Faster is cheaper. The longer the program duration, the higher the costs. Ways to reduce program duration include designing programs with maximum use of

People at ESA

"All this stuff is decided behind closed doors, with no dialogue and no real outline of what the goals are," said one ESA employee. "People around here—many of whom work very hard—feel like they have not been treated with respect."

Quoted by Peter B. Selding in "Tension Level High at ESA" Space News—Aug. 5-11, 1996, p. 3.

already developed components, and verifying advanced technology items before the contract is initiated to minimize schedule stretchouts.

Lack of Standard and Available Hardware, Software and Designs— Unfortunately, no standard equipment program exists today, and no efforts are currently underway to identify and catalog qualified off-the-shelf equipment and software, although their importance to NASA's faster, better, cheaper effort is well recognized. There may be no more effective way to control cost than to use already developed equipment and software. New technology development, when necessary, should be conducted in separate, non-program, experimental programs. As Tischler suggested, impressive cost savings would come from payloads assembled from off-the-shelf equipment.

The Size of the Staff—Time and again studies revealed that to reduce cost organizations should reduce the size of project staffs. It was common knowledge that there was a direct correlation between the number of government employees and the resultant size of the contractor's work force. Low told Tischler: "On management approaches, the best guideline I could give is to run a 'tight ship,' one that is 'lean and mean.' This means that at each level in the organization— Headquarters, Centers, contractors, subcontractors—there are only sufficient people so that they are all 'doers' and there are no 'hangers on.'" Kurt Debus, the KSC Center Director, observed that cutting back on people may be the only real key to reducing costs.

Excessive Documentation—It has been estimated that program documentation can cost as much as 15% of total program cost. If NASA were to back away from partnering with its contractors, it would not need the amount or level of contractor-generated data presently called for and could use more of the information already generated by the contractors for their internal management, greatly reducing program, personnel and publication costs.

Requirements—Poorly defined requirements are directly tied to buy-ins and poor definition, and they are major elements of cost overruns. As more than one NASA employee observed, the greatest potential for low cost programs is in a complete and comprehensive definition of requirements prior to initiation of a program, accompanied by a control program to guard against requirement creep.

Complexity—Today, management's red badge of courage is complexity. Managers seek out the challenge of managing complex programs. Much of this complexity intended for large space projects is often applied to simpler tasks. The process is stultifying, with a corresponding direct effect on costs. Consider the following example. Freeman Dyson of Princeton's Institute for Advanced Study, writing in the Forum section of "Issues in Science and Technology," tells the story of a colleague, Bob Dicke of Princeton's physics department, proposing an experiment to NASA. Dicke wanted to place reflectors on the moon, then fire lasers at them to measure the return time and hopefully gain a better understanding of the moon's motion. Dicke would buy the reflectors off-the-shelf from a science supply house and have his department's shop prepare the necessary trays and stands. His cost estimate to NASA was $5,000 per unit. NASA bought the proposal but insisted on contracting out the fabrication to an aerospace contractor who would follow

NASA processes, procedures and specifications. NASA ultimately paid $3 million per unit.[7]

A Low Cost Program Described

What, then, would an ideal low cost program look like, given the findings of the LCSO? First and foremost, this "new way of doing business" would have better respect for the work force. Time and again, the Low Cost Systems Office of the mid-1970s found that people are the single most important element of a truly low cost program. A small, competent, highly motivated and fully rewarded staff, directed by an experienced manager with clear and concise goals, is the first step in achieving genuine cost control.

The implications of such an effort are enormous. The Low Cost Systems Office found that small, flat project organizations with horizontal communications work best, and most efficiently. Such a lean organization, ever mindful of the past, seeks to learn from previous projects, both successes and failures. Lessons learned are the best source of unique management experiences.

This carefully selected team will have before them a set of clearly stated requirements that will be fixed for the lifetime of the program with changes allowed only to ensure success. An adequate definition will be performed to fully understand the scope, complexity and expected costs and to allow for the trade off of cost against performance, if that becomes necessary. The past quarter century of "buying-in," a practice deplored by the Low Cost Systems Office, shows no signs of abating. This must be stopped if costs are to be brought under control. Severe sanctions are needed.

This low cost program will not try to reinvent the wheel, as George Low was fond of saying. Instead, whenever possible, the team will use previously developed designs, hardware, software, documentation and proven management tools. It is always cheaper to assemble than develop. Old LCSO terms like "off the shelf," "standard equipment" and "standard practices" will once again become part of the nomenclature of the project planning team.

Finally, such a low cost program would not waste time. One firm finding of the Low Cost Systems Office was that "haste saves money." Any number of techniques will speed up a program without "making waste." A few of them are mentioned earlier, such as thorough definition, fixed requirements, simple interfaces and doing only what is necessary. A faster project is not only cheaper, but usually it is better because it is simpler, streamlined.

All this was known in the mid-1970's. Unfortunately, subsequent generations simply forgot it and today government is busy reinventing it as faster, better, cheaper.

Footnotes

[1] Richard Luecke, Scuttle Your Ships Before Advancing: And Other Lessons From History on Leadership and Change for Today's Managers. Oxford University Press, New York, 1994.

[2] Gilbert Roth, "Lessons Learned and Their Application to Program Development and Cultural Issues," Space Transportation Propulsion Technology Symposium, Pennsylvania State University, June 1990.

[3] NASA, Lexicon Version 1.1, November 1991, p. 20.

[4] "Planning the Future in Space: The Continuing Search," a briefing paper of the NASA Advisory Panel of the National Academy of Public Administration, July 14, 1989.

[5] Frederick J. Haskin, The Panama Canal. Doubleday, Page and Company, 1913.

[6] Donald P. Hearth, "Project Management in NASA: 1980 and Today," Issues in NASA Program and Project Management (NASA SP-6101-04), Spring 1991.

[7] Freeman J. Dyson, "The Final Frontier?" in Issues in Science and Technology, VIII, No. 1 (Fall 1991), p. 6.

Chapter 15

A New Way of Doing Business in Space

In 1998, NASA will be 40 years old, the average life expectancy of a Fortune 500 company.[1] During this time NASA has much to be proud of. Indeed, the NASA of earlier years was held up as a shining example of what the U.S. Government could accomplish. But today's NASA is an artifact of a time gone by.

NASA has survived the post-Apollo era without the changes that George Low thought necessary for survival. Apollo was costly, but it was incredibly successful. Post-Apollo NASA has also been costly, but not as successful. During the past decade, friends and foes alike have questioned NASA's ability to manage complex projects. In a report entitled *Space Missions Require Substantially More Funding Than Initially Estimated*, project after project reviewed by the General Accounting Office has been long delayed, cancelled or ended in failure.[2] Billions of research and development dollars have been lost on these unsuccessful projects. When civil service salaries, travel, transportation and other expenses, including new and modified facilities, are added, the total costs approach staggering proportions.

Also damaging is the belief by some that with the tens of billions spent on the Space Station program since its approval in 1984, some version of a U.S. laboratory would be orbiting the Earth today if NASA and the Congress had not been "micro-managing" its development. Is the NASA partnering management process now so expensive as to preclude its use on new space programs? To examine that question, it is necessary to draw sound comparisons and contrasts between NASA's Apollo-era management style and one that may be better suited for the future.

The Apollo management model, along with its carefully crafted large organization, was not intended as the model to take NASA into the future. James Webb, for all his brilliance and savvy, had no firm plans for the post-Apollo NASA. Perhaps he figured that the NASA management structure and organization would be dismantled after Apollo and future government participation in civil space would require a new model. However, with no blueprints for its future, NASA continued to cling to its past. The Apollo-style organization is long overdue for change, as well as the role of government and its agent, NASA, in civil space programs.

This scenario was recognized decades ago. In July of 1971 the senior scientist in NASA's Office of DoD and Interagency Affairs, Milton Rosen, sent a note to Fletcher, Low and others, concerning a proposal he had received from a member

Although "Gedanken" appears on the original memo, the proper word appears to be "Gedenken." "Danken" is translated "to thank."

of the Aeronautics and Space Engineering Board. The proposal, "A Gedanken (thought) Experiment," concerned NASA's yet to be developed Space Shuttle. Portions of this proposal are included here in order to show that even among aerospace advocates, NASA's days as an institution appeared numbered.

A "Gedanken Experiment"
About NASA and the STS (or the "Shuttle")

Suppose NASA as it exists now were to vanish, and in its stead other entities would carry out NASA's present functions, in the following manner and framework:

The various NASA Centers and laboratories would continue their various kinds of researches and developments, such as:

Lewis—would work on advanced propulsion systems

Ames—on hypersonic flow problems

Goddard—on guidance and control

JPL—on unmanned planetary exploration

Huntsville—on large propulsion systems

Langley—on STOL and VTOL technology

Houston—on manned space flight problems and so on, some of these researches being funded by an (enlarged) National Science Foundation

A new Corporation would be set up, by the Government, but publicly owned and partially publicly controlled (in the manner of COMSAT) which would go into the business of space transportation for the nation as a commercial venture.

In addition to the capital raised by the initial stock offering, its assets would be: the family of launch vehicles (developed at Government expense), the launch facilities at the Cape, and the other telemetry and communications facilities around the world now run by NASA, and those portions of other NASA facilities directly concerned with the development, design, engineering, testing and contracting of launch systems. Its customers would be: the various government agencies which require launch services, such as ESTA, DOT, NSF and perhaps DoD; also COMSAT, AT&T and whoever else needs satellites and spacecraft launched, including customers abroad.

The question I should like answered is this: What kind of space transportation system would this corporation—let us call it the "Space Transport Corporation"—develop in the course of the next two decades?

Is it obvious that the STC would immediately begin with the Space Shuttle-plus-Space Tug concept, on the scale now being proposed by NASA?

I think not. In fact, I believe the approach of the STC would be much more conservative, developing and adapting the shuttle vehicle capacities to fit the absolutely minimal projected traffic, and using the presently available expendable launch vehicles to take care of any extra unforeseen traffic, and perhaps developing some new low-cost expendable ones as well.

When I got this far with my "thought experiment," I was tempted to think: why not consider the whole question more seriously?

NASA, *in its present form*, was designed to be the means—splendidly successful though it was—of carrying out the Apollo mission. In its present form it may no longer be the best means for achieving the avowed purpose of opening up the new frontier of space.

Having used the Apollo approach once, NASA appears to believe that this is the best and only way of doing business; NASA does not appear to realize Apollo was an entirely political act, conceived as such—and entirely successful in the process—and that the shuttle is an entirely different matter.

Surely, "The purpose of government is to do things for society which cannot be done by individual or combined private efforts."

The business of carrying payloads into and out of space is something that should be done by "business," or if it is done by government, it should be done as business-like as possible. The Apollo style is far removed from business-like.

Is it really too far-fetched to imagine a company, -STC-, its personnel largely from NASA and from the aerospace industry, incorporated by Act of Congress, to carry on the space transportation business of this country?

The reconstituted NASA, or its parts, would carry on the research and development much as it is done now, funded by Congress or through NSF, as applicable. The proposed Space Transport Corporation would finance its operations on the open market, and the public would be given a chance to participate in this new and exciting venture. The opening-up of the space frontier:

What would be the advantages?

It would remove a large part of the space activities from the political scene. No longer would Congress have to immerse and inject itself into highly technical matters. No longer would "space" have to compete with "urban decay" or "environmental pollution." Space transport would be developed under the influence of the demand and the technological opportunities, at a rate which made economic and technical sense, and which presumably coincide with the greatest social benefit to the nation.[3]

The proposal created a stir in NASA's executive suite. Fletcher, Low and von Braun were concerned that the author recognized the downgraded Shuttle not as a low cost launch system, but as a "Save the Agency" program. They rationalized , however, that NASA would eventually be vindicated when some future Administration finally approved a Space Station with the Shuttle as its transportation system.

In March 1975, Richard Chapman of the National Academy of Public Administration and author of *Project Management in NASA: The System and Men*, wrote to the author:

It is clear that the management structure built in the sixties and generally continuing down today was a structure more related to managing rather large scale projects while simultaneously being rather deficient in developing, enunciating, or examining long range Agency goals. As we urged in our project manager report, this type of structural system has a built-in lack of organizational continuity. Concentration upon time-limited tasks and operational relationships built upon personal ties appears to weaken an agency when those tasks are completed or curtailed and when key people leave. This certainly has come to pass with a vengeance in NASA.[4]

In the intervening years there have been numerous attempts, both internal and external, to "study" NASA's management and organization, but none produced significant change from Webb's "partnership" model. In almost a quarter century of hanging on, NASA has, to its credit, made significant and occasionally outstanding contributions to science, engineering and technology. It has had a beneficial effect on lifting the human spirit and has made life on Earth better and more meaningful for all its inhabitants, but it is still an artifact of another time in this nation's history.

A New Paradigm

"America needs space to grow" was a popular NASA slogan of the 1980's. It seemed to reflect the realization that the nation's civil space program needed to transition from its past focus on national security to one of economic opportunity. Unfortunately, it was a slogan, not a sign of real change. Government continues to dominate civil space and to invest in itself at the expense of innovative nongovernmental efforts, a paradigm totally inappropriate in times of intense international economic competition, scarce resources, government downsizing and re-engineering. The results of this behavior are devastating. The U.S. has lost leadership in any number of civil space categories. For instance, in the past 20 years it is estimated that the U.S. has lost 66% of its commercial launch markets and 24% of its commercial satellite markets.[5] These numbers will most likely continue to increase as the competition grows in number and competence, a strong indication that neither the nationalized aerospace industry nor its government partner made the necessary investment in commercial space opportunities.

In the quarter of a century since the flight of Apollo 17, many events have occurred that warrant or perhaps demand a change in the manner the government involves itself in civil space. Foremost among these is the evaporation of the very threat NASA was founded to counter. The Cold War is over. Civil space is due a peace dividend. In addition, the general shift from the public to private sectors, and the necessity of doing more with less in times of fiscal constraint, demand change in the government-dominated, overly costly civil space program. The issues of Earth sustainability and global environments also seem to be beyond the ability of today's government to accommodate. A true space hero, Dr. George Mueller, who was the NASA Associate Administrator for Manned Space Flight from 1963 to 1969, speaking at one of many Apollo 11 25th anniversary celebrations, reflects the thoughts of many:

> Throughout this year there are and have been celebrations occurring throughout our country and in many foreign countries, celebrating the first venture of mankind to another planet, now 25 years ago. Yet it is a sad commentary on our political system that the "giant leap" for all mankind has been frozen by inaction for all that time. Worse yet, by failing to establish a purpose, a goal and the organization to accomplish it, we have spent over that time all the money that would have been required to establish a colony, for example, on the moon.[6]

Mueller indicts the government's civil space policy of fostering organizational survival at the expense of exploration and exploitation. Radical change is needed to end the old organization and to build an exciting and economically rewarding new civil space program and infrastructure, with an emphasis on content, not organizational survival.

NASA's Closed Shop

"Before we all get swept away in the single-stage-to-orbit (SSTO) X-33 vehicle, we should pause to remember that the X-33 is just another NASA vehicle, that NASA continues to hold a monopoly on manned space-flight and that NASA is in no way committed to the development of a free enterprise space transportation industry in the United States."

Mark Gall,
in a Letter to the Editor
Space News—
Aug. 26-Sept. 1, 1996.

The promise of space has yet to be fulfilled. Until low cost transportation and orbital operations are available to new customers, especially the groups Low targeted, the exploitation of this potentially important economic opportunity is unnecessarily delayed. Unfortunately, the current and past government role is a powerful deterrent to the exploitation of space. The only users of the space so far have been the quasi-nationalized aerospace companies under direct contract to NASA and other government agencies. Few new customers, especially the entrepreneurs, can afford access to space. Low's vision of inexpensive transportation for the "cheap payloaders" is still thwarted by government bureaucracies, working in their own best interest, at the virtual exclusion of new participants. For all practical purposes, it's a closed shop. However, innovation frequently comes from these small independent businesses, not huge corporate conglomerates or government bureaus. These independent pioneers are vital to a new era in space, because as a group they will most likely identify and exploit the new opportunities of space. They do not require hundreds of millions of dollars annually but only modest start-up costs, cheap space transportation and access to tools, methods and processes developed to support other programs.

Two major changes are needed to revitalize space activities. First, and most important, is cheap access to space. This is an absolute imperative for future progress in space. The second is a radical change in the current space management paradigm, so as to reduce cost and stimulate new users with the necessary private investment capital. Vested interests accustomed to the old way of doing business may be threatened, but the lack of progress during the past 25 years defeats their arguments for continuation of the status quo. Machiavelli reminds us, "There is no more delicate matter to take in hand nor more dangerous to conduct, nor more doubtful in its success, than to set up as a leader in the introduction of changes. For he who innovates will have for his enemies all those who are well off under the existing order of things, and only lukewarm supporters in those who might be better off under the new!"[7] There are a large number of those well off under the existing launch and management arrangements.

In any reexamination of the proper role of government in the future of space, it is helpful to categorize the civil space program into the following categories: Research, Exploration, Exploitation and Operations.

Research

NACA, the renowned aeronautical research organization, founded in 1915, was the core of the new NASA. NACA's product—the research report—was respected by the world's aeronautical community, along with its technology transfer mechanism enabled by advisory committees composed in part of their customers. Few questioned the government's role in the conduct of aeronautical research and, later, space research. This historical role should be continued by NASA and should be strengthened with new research and computational facilities and capabilities, along with enhanced technology transfer mechanisms and increased cooperative endeavors.

212

Exploration

To continue the astonishing progress made over the past decades in exploring the Earth and its environs, the moon, planets and cosmos, the government should continue to fund exploration programs, possibly through the National Science Foundation. The NSF, with its academic/industry customers, would play a much more influential role in identifying, justifying, budgeting and overseeing any new exploration initiatives. The exploration emphasis would be on the collection of data rather than the building of hardware. Innovative procurement techniques structured to reward performance would be employed. NASA's research capabilities would continue to support the exploration efforts.

Exploitation

The organizations and individuals Low believed NASA had to reach out to in order to launch a new era in space—the cheap payloaders, space pioneers, entrepreneurs, academicians, small science and commercial advocates and venture capitalists—would all be accommodated in programs fully or partially funded by the government, depending on cost sharing agreements. This group will best exploit the promise of space and initially lead the commercial effort most effectively. The government can have a partnering role with this group by funding government/industry or otherwise supportive proposals that promise real or potential economic payoff. There are plenty of successful cost sharing precedents to learn from.

Operations

Typical space operations would include launch services, Space Station ground and on-orbit services, weather satellites and Earth monitoring satellite systems. These operations should be dominated by private sector companies. The government would buy these services as would industry and other users. Operations should be self-supporting through the sale of services.

Where Do We Go From Here?

The first century of space exploration rapidly comes to a close. In 1898 Konstantin Edwordovitch Tsiolkovsky, a Russian school teacher, published a formula for a rocket device to demonstrate by calculation that humans could escape the gravitational pull of the Earth and travel in space. This marks the beginning of the space age.

The year 1998 is also the 40th anniversary of the founding of NASA. An international celebration to commemorate the achievements of these first hundred years could redirect the future of space flight as well as honor those who came before us. And there is much to celebrate. The accomplishments made within the span of just two or three generations are nothing short of spectacular.

NRC Urges Commercial Use of Station

The National Research Council's 76-page report, "Engineering Research and Technology Development on the Space Station," should be considered only a first step down the road to privatizing the space station . . .

The NRC report concluded that NASA's plans to date for commercial use of the station not only overlook engineering research and development, but also "seem more likely to support projects that require subsidies for the foreseeable future than to result in commercial use of the station." It also states that, "NASA must stop trying to pick winners and use market-based mechanisms to select commercial experiments instead."

by Anne Eiseli in Space News— June 24-30, 1996, p. 4.

A Narrower Space Agency Authority

"It is noticeable, however, that in many parts of the world, the authority of national or international space agencies is narrower than it used to be. In most regions there is an increasing number of ministries, departments and agencies that have to be consulted, and this consultation is by no means simply a matter of form."

Roy Gibson, Former ESA Director-General in "Private Sector Drives Development" Space News— May 27-June 2, 1996, p. 19.

Built on the liberating theories of Tsiolkovsky, inspired by the daring flights of the Wright brothers and assured by the pioneering work of Robert Goddard, earthlings have explored virtually the entire solar system and far beyond. We have penetrated the astonishing mysteries of the Universe and made access to outer space routine, if not cheap. If past is prologue, the future is indeed bright.

Any celebration of these past accomplishments ought also to focus on the future. In a series of debates open to the public and the media, the character, scope and nature of this nation's future space initiatives will be determined once and for all. The involvement of the public is an important and necessary ingredient due to the fact that people will support whatever they help create.

The outcome could be a dynamic new federal space agenda on which to build new dynamic activities. Another major outcome of such celebrations could be the coming to closure of the first space age and the proper launching of a second century of space flight with even greater hopes and expectations for the betterment of humanity. After all, that's the ultimate purpose of space flight. While space exploration is often seen as an end in itself, ultimately it can and should lead to living better, richer lives and to the enhancement of our human destiny in the cosmos.

Footnotes

[1] See Peter M. Senge, The Fifth Discipline: The Art and Practice of the Learning Organization. New York, Doubleday, 1990.

[2] General Accounting Office, "Space Missions Require Substantially More Funding Than Initially Estimated." Report No. 93-97, December 1992.

[3] Memo to A/Administrator from Milton Rosen, July 16, 1971.

[4] Memo to Frank Hoban from Richard L. Chapman, National Academy of Public Administration, March 3, 1975.

[5] Industry Interface Group, A Team Approach to Global Space Commerce, NASA Strategic Avionics Technology and the American Institute of Aeronautics and Astronautics, Arlington, Virginia, November 1993.

[6] George Mueller, "On the 25th Anniversary of Apollo 11," an address to the NASA Headquarters Alumni gathering on October 5, 1994.

[7] Niccolò Machiavelli, The Prince. Translated by Hill Thompson, New York, The Heritage Press, 1954.

List of Acronyms

AAP	Apollo Applications Program
ACE	Acquisition Cost Evaluation
AIA	Aerospace Industries Association
AIAA	American Institute of Aeronautics and Astronautics
ARC	Ames Research Center
ARD	Atmospheric Re-entry Demonstrator
ASEB	Aeronautics and Space Engineering Board
ASME	American Society of Mechanical Engineers
ATS	Advanced Technology Satellite
CASH	Catalog of Available and Standard Hardware
COMSAT	Communications Satellite Corp.
CPAF	Cost Plus Award Fee
CPFF	Cost Plus Fixed Fee
CPIF	Cost Plus Incentive Fee
CV	Convair
DAD	Dual Air Density
DFRC	Dryden Flight Research Center
DoD	Department of Defense
DOT	Department of Transportation
DSMC	Defense Systems Management College
EIA	Electronics Industry Association
ESA	European Space Agency
FP	Fixed price
FY	Fiscal year
GFE	Government-furnished equipment
GFP	Government-furnished property
GREMEX	Goddard Research Engineering Management Experience
GSFC	Goddard Space Flight Center

HCMM	Heat Capacity Mapping Mission
HEAO	High Energy Astronomy Observatory
IRAS	Infra-red Astronomy Satellite
IUS	Inertial Upper Stage
JPL	Jet Propulsion Laboratory
JSC	Johnson Space Center
KSC	Kennedy Space Center
LaRC	Langley Research Center
LCC	Life cycle costing
LCSD	Low Cost Systems Division
LCSO	Low Cost Systems Office
LeRC	Lewis Research Center
LST	Large Space Telescope
MMS	Multi-Mission Spacecraft
MOL	Manned Orbital Laboratory
MSC	Manned Spacecraft Center
MSFC	Marshall Space Flight Center
MVM	Mariner Venus Mercury (project)
NACA	National Advisory Committee for Aeronautics
NASA	National Aeronautics and Space Administration
NERVA	Nuclear engine for rocket vehicle applications
NIH	"Not invented here"
NMI	NASA Management Instruction
NRC	National Research Council
NSF	National Science Foundation
OAO	Orbiting Astronomical Observatory
OART	Office of Advanced Research and Technology

OAST	Office of Aeronautics and Space Technology		**SAGE**	Stratospheric Aerosol and Gas Experiment
OJT	On the job training		**SAS**	Small Astronomy Satellite
OMB	Office of Management and Budget		**SEASAT**	Sea Satellite
			SEB	Source Evaluation Board
OMSF	Office of Manned Space Flight		**SR&T**	Supporting Research and Technology
OSO	Orbiting Solar Observatory			
OSSA	Office of Space Science and Applications		**SSTO**	Single-stage-to-orbit
			"STC"	"Space Transport Corporation"
			STG	Space Task Group
PMSEP	Project Management Shared Experience Program		**STOL**	Short Takeoff and Landing
			STP	Space Test Program
PRC	Planning Research Corporation			
PSAC	President's Science Advisory Committee		**TIROS**	Television and Infrared Observation Satellite
			TQM	Total quality management
R&D	Research and Development		**TRW**	Thompson-Ramo-Woolridge
R&QA	Reliability and Quality Assurance		**UARS**	Upper Atmosphere Research Satellite
RCA	Radio Corporation of America			
RFP	Request for Proposal			
RPI	Rensselaer Polytechnic Institute		**VTOL**	Vertical Takeoff and Landing
SAE	Society of Automotive Engineers		**WBS**	Work breakdown structure

Index

Life (mag.), 3
Life cycle costing (LCC), 90
Lockheed Missiles and Space Company, 87–89, 117, 157
Low, David, 13
Low, George, 5, 11, 13–23, 25–42, 45–46, 51, 69, 78, 93, 94, 153, 203, 207, 210, 211; and Apollo Program, 5, 14, 28, 59, 191, 193, 196; and ASEB, 136; background of, 13–14, 191; and Berlew report, 155; and Cooper, 193; and cost-cutting, 25–39, 57–59, 99, 119, 158, 177–79, 181–83, 187, 188, 192–94, 197, 205, 212 (*see also* Low Cost Systems Office); departure of, 125, 128, 155–56, 165, 191, 193, 198; and Gray, 58, 125, 155, 192; internal criticisms of, 151; and Johnson, 125, 138; and LCSO, 53, 54, 131–32; legacy of, 193–94; and McCurdy, 10; and Mandell, 179; and MMS, 70; and Muinch, 61, 193; and Petrone, 85; and Project ACE, 90; and Space Shuttle, 191, 192; as success-obsessed, 150, 186; and Tischler, 33, 34, 36, 37, 46, 49, 53, 58, 59, 192, 195, 204; training endorsed by, 139
Low Cost Systems Board, 131–33. *See also* Gray, Ed
Low Cost Systems Division (LCSD), 127
Low Cost Systems Office (LCSO), 16, 53–55, 90, 126, 138, 139, 142, 165–75, 184, 190, 193–95, 197, 198, 200, 203, 205; cost-cutting efforts of, 79–82, 93, 99, 131–34, 145–47; as "cultural enemy," 178, 181; demise of, 73, 170, 198; evaluation of, 171–74; failures of, 189–91, 195; fighting for life, 170; hardware focused by, 57; lessons learned from, 196–205; payloads focused by, 157–58, 161; project managers focused by, 99–129; reorganization of, 170; resistance to, 185–89, 193, 197, 198; Sonnemann investigation of, 166–67; and standardization, 58, 60, 61, 63, 65–70, 73, 170, 171; studies of, 92; successes of, 186,

193, 198. *See also* Gray, Ed; Johnson, Vince; Low, George; Muinch, George; Tischler, Del; Workshops, project management
Low cost workshops. *See* Workshops, project management
Luecke, Richard, 195
Lunar landers, 112
Lunar Orbiter project, 104, 110
Lunar rover, 5–6
Lundin, Bruce, 35
Lunney, Glynn, 96, 139

McCurdy, Howard E., 3, 181
McCurdy, Richard, 5, 10
McDonnell Douglas Astronautics Company, 120
McGee, Frank, 50
Machiavelli, Niccolò, 212
Management, NASA, 83–84, 99–129; LCSO focus on, 79, 82; systems, 135–36; unqualified, 202. *See also* Apollo Program, management style of; Contracts; NASA project managers; Purchasing
Mandell, Humboldt C., 178–81
Manned Flight Awareness program, 145
Manned Orbital Laboratory (MOL) project, 1
Manned Spacecraft Center, 14, 135
Manned Space Flight Program, 14
Mariner Mars project, 117–18
Mariner project, 21, 32, 34, 80, 124
Mariner Venus Mercury (MVM) project, 80, 101, 110, 120
Mariner-Venus project, 99
Mark, Hans, 77
Mars (planet), 2, 104, 112, 117; man to, 179
Marshall Space Flight Center (MSFC), 37, 67, 108, 115, 145, 161, 166, 196
Mars Observer, 173
Martin, Jim, 36
Mercury (planet), 101. *See also* Mariner Venus Mercury project
Mercury Awareness program, 145
Mercury program, 14, 145, 178
Metzenbaum, Sen. Howard, 9–10
Metzger, Sid, 99, 110–13, 118, 120
"Mighty" Low Cost (cartoon character), 147
Minnesota, University of, 115

Missiles, as launch vehicles, 145
Mission Office Program, 43
Models, 172
Modularity, 71, 73–74, 87, 89, 169. *See also* Standardization
Money, as motivator, 146. *See also* Costs, NASA and
Moon: colonization of, 211; Dicke designs on, 204; man on, 1–6, 13, 78, 112, 135, 179, 191, 195, 211; mapping of, 198; scientific data from, 4. *See also* Lunar Orbiter project
Motivation: contractor, 119, 133, 146; expectancy theory of, 189; financial, 146; of NASA employees, 123, 133, 152; Tischler focus on, 47
Motivation programs, 145–48, 152
Motors, rocket, 14
Mueller, George, 211
Muinch, George, 60–62, 71, 131–32, 193
Multi-Mission Spacecraft (MMS), 70–74, 195
Murphy, Pat, 61
Myers, Dale, 32, 135, 136

Nadler, E. E., 189
NASA (National Aeronautics and Space Administration): accomplishments of, 7, 8, 210 (*see also* Moon, man on); advantages to public of, 7; and Air Force bidding procedures contrasted, 37; birth of, 41, 145; bureaucratization of, 3, 4, 59, 170, 177, 178, 181–85; business procedures of, 77–98, 169, 178, 180, 187, 195, 205, 207, 209, 212 (*see also* Aerospace industry, NASA and); and COMSAT contrasted, 36, 37, 51; contractors and (*see* Aerospace industry, NASA and); cost considerations of (*see* Costs, NASA and); "culture" of, 60, 145–56, 177–81, 183–85, 187, 195, 197, 198; Cutter on, 181; decline of, 1–5, 211; and DoD contrasted, 29–30; failures of, 207 (*see also* Apollo Program, tragedy-stricken; Space station); federal constraints on, 145–46, 150; Fletcher boosting of, 7–8;